Collins

GCSE Maths
2 tier-higher
for AQA B
TEACHER PACK

GREG BYRD

LYNN BYRD

PHIL DUXBURY

William Collins' dream of knowledge for all began with the publication of his first book in 1819. A self-educated mill worker, he not only enriched millions of lives, but also founded a flourishing publishing house. Today, staying true to this spirit, Collins books are packed with inspiration, innovation and a practical expertise. They place you at the centre of a world of possibility and give you exactly what you need to explore it.

Collins. Freedom to teach.

Published by Collins
An imprint of HarperCollins*Publishers*
77–85 Fulham Palace Road
Hammersmith
London
W6 8JB

Browse the complete Collins catalogue at
www.collinseducation.com

© HarperCollins*Publishers* Limited 2006

10 9 8 7 6 5 4 3 2

ISBN-13 978-0-00-721574-4
ISBN-10 0-00-721574-6

British Library Cataloguing in Publication Data. A Catalogue record for this publication is available from the British Library.

Commissioned by Marie Taylor and Vicky Butt

Publishing Manager Michael Cotter

Project managed by Nicola Tidman

Edited by Marian Bond and Paul Sterner

Proofread by Amanda Whyte

Internal design by Gray Publishing

Cover design by JPD

Cover illustration by Andy Parker

Page make-up by TECHSET Composition Ltd

Illustrations by TECHSET Composition Ltd

Production by Natasha Buckland

Printed and bound by Martins the Printers, Berwick upon Tweed

Acknowledgements
Whilst every effort has been made to trace the copyright holders, in cases where this has been unsuccessful or if any have inadvertently been overlooked, the Publishers will be pleased to make the necessary arrangements at the first opportunity.

Contents

Introduction

Welcome to Collins GCSE Maths!

Collins GCSE Maths uses a stimulating approach to Maths that really appeals to pupils, and features straightforward guidance on grade performance for clear progression. Written by experienced teachers and examiners, the series provides perfect coverage of the new 2 tier AQA GCSE Modular B Maths Specification.

This Teacher Pack has been developed to make teaching GCSE Maths much easier. It not only supports the Collins GCSE Maths textbook, but also provides a wealth of resources such as homework activities, worked examples and extra prior knowledge "Check-in" tests.

Developed by the team behind the highly popular and successful Maths Frameworking series, Collins GCSE Maths is the easiest way to achieve success in 2 tier Mathematics.

Teacher Pack

Chapter Overviews
Each chapter commences with a Chapter Overview. These provide a summary of the work pupils will encounter, and offer context for the key mathematical ideas – this will help explain "why is this topic useful to us?"

Links to the specification are given, followed by a Route mapping grid through the Exercises to indicate the level of work pupils will meet. The grid will show, for example, that questions 1–6 in Exercise A are set at around Grade F difficulty level.

In addition, answers to the diagnostic Check-in test are provided.

Check-in tests
Each chapter has a short diagnostic test which can be copied and distributed to the class, before main teaching of the topic begins. This will help quickly establish if pupils have the appropriate level of understanding to tackle the topic.

Lesson plans
Each section in the textbook is supported by a Lesson plan. The Lesson plan provides:

- Key words, Learning objectives and Links to other Pack resources for each section.
- Engaging oral and mental starter activities to involve the whole class. These are designed to work with minimal specialised equipment, although of course the use of whiteboards, digital projection, OHPs, target boards, counting sticks, etc. can make the activities easier to present and more accessible to pupils.
- Main teaching points to help lead pupils into exercise questions.
- Notes on common mistakes that pupils may make.
- Notes on differentiation – advising how to direct pupils of varying ability to the work that best suits them.
- Plenary guidance to complete the structured lesson.

Homework answers
The Teacher Pack provides Homework activities for every section. The answers to these activities are provided towards the back of the Pack.

Textbook Examination questions answers
Answers to the exam practice questions in the textbook are provided towards the back of the Pack.

Note
The new Specification B assumes that pupils taking Modules 1 and 3 have some basic understanding of other assessment objectives.

For example, in Module 1 (Data handling) and Module 3 (Number) pupils should know:

- that letters can represent unknown quantities
- how to substitute numbers into simple algebraic expressions.

Also in Module 1 (Data handling) pupils should know:

- how to cancel fractions to their simplest form
- that percentage means 'out of a hundred'
- the meaning of a simple ratio.

Also in Module 3 (Number) pupils should know:

- how to combine like terms and manipulate simple algebraic expressions
- how to solve simple equations.

This means that there will be a new style of question. These are flagged with the Exclamation Mark icon in the Pupil Book. These questions already appear in Specification A.

Teacher Pack CD-ROM

All of the printed materials in the Teacher Pack are also supplied on the CD-ROM, to allow customisation and easy access. The structure of the CD-ROM matches the structure of the textbook, and by navigating to the required section, you will find the Chapter Overviews, Check-in tests and Lesson plans in Adobe Acrobat PDF form.

In addition, the CD-ROM also features:

Homework activities
Each section in the textbook is accompanied by a homework activity. These can be printed out and distributed to the class. Note that answers to these activities are listed at the back of the Pack.

Worked examples
Most of the sections in the textbook are accompanied by at least one Worked example on the CD-ROM. Use these to provide extra support if needed to help pupils grasp key ideas.

Bonus content online

In order to further assist you in your teaching, Collins will provide ongoing updates on a dedicated section of its website. For access to this FREE bonus content, such as editable MS Word™ versions of the Lesson plans, go to:

www.collinseducation.com/gcsemaths/mathsbonuscontent

This area of the website can also be accessed by clicking through the hyperlink on the CD-ROM.

We do hope you enjoy using Collins GCSE Maths, and wish you good luck in your teaching!

Greg Byrd, Lynn Byrd, Phil Duxbury, Will Ferguson and Gillian Read

Overview

1.1 Averages
1.2 Frequency tables
1.3 Grouped data
1.4 Frequency diagrams
1.5 Histograms with bars of unequal width
1.6 Moving averages
1.7 Surveys
1.8 Social statistics
1.9 Sampling

This chapter covers the basic statistical topics required at this level. Measures of location of discrete and continuous data are included (excluding the median estimate of continuous data), along with the displaying of data in the form of frequency polygons and histograms. The four final sections comprise two distinct topics on moving averages, and surveys, social statistics and sampling.

Context

Averages are of course used in a huge variety of everyday situations, from schools (test marks, attendance figures) and shops (daily/weekly takings, annual turnovers, profit margins) to finance (planning and making forecasts) and sport (batting averages, attendance figures). Share prices and changes in currency can initially appear erratic, but their general trends can be observed through the practical use of moving averages. These also have their use in actuarial work.

The sections on surveys and sampling at the end of the chapter are also particularly important, given their likely prominence in regard to statistical coursework tasks. Pupils may well have to design and conduct a survey or questionnaire, and explain, for example, the type of sampling they have chosen (and why) when writing up their task.

AQA B references

AO4 Handling data: Processing and representing data

1.1–1.3   4.4e "... calculate the mean for large data sets with grouped data"
1.4–1.5   4.4a "draw and produce, using paper and ICT, pie charts for categorical data, and diagrams for continuous data, including line graphs (time series), scatter graphs, frequency diagrams, stem-and-leaf diagrams, cumulative frequency tables and diagrams, box plots and histograms for grouped continuous data"
1.6   4.4f "calculate an appropriate moving average"

AO4 Handling data: Interpreting and discussing results

1.1–1.3   4.5d "compare distributions and make inferences, using shapes of distributions and measures of average and spread ..."
1.6   4.5b "interpret a wide range of graphs and diagrams and draw conclusions; identify seasonality and trends in time series"
1.8   4.5j "interpret social statistics including index numbers; time series; and survey data"

AO4 Handling data: Specifying the problem and planning

1.7–1.9   4.2d "...select and justify a sampling scheme and a method to investigate a population, including random and stratified sampling"
          4.2e "design an experiment or survey; decide what primary and secondary data to use"

Route mapping

Exercise	D	C	B	A	A*
A	1–7	8–10			
B	1–5	6–7			
C		all			
D	1–2	3–5			
E				1–4	5–8
F			1–2	3–6	
G	1–4	5			
H	1–2	3–6			
I	1–4		5	7	
J	1	2–5		6	

Answers to diagnostic Check-in test

1 a 4   **b** 130

2 a 8   **b** 3

3 a 9   **b** 9.5   **c** 3

4 a 8   **b** 400

5 a 39.125   **b** 49.125

6 a i 61.3   **ii** 64.8   **b** 63.1

7 15

    Module 1: Data handling

1 Find the mean of these sets of numbers.
   a 6, 0, 4, 4, 5, 8, 3, 2, 2, 6
   b 110, 150, 100, 120, 170

2 Find the mode of these sets of numbers.
   a 2, 5, 6, 6, 7, 7, 8, 8, 8
   b 6, 4, 4, 3, 6, 7, 9, 9, 1, 9, 2, 3, 5, 3, 8, 3

3 Find the median of these sets of numbers.
   a 6, 7, 9, 9, 10
   b 6, 7, 9, 10, 10, 20
   c 5, 7, 1, 0, 2, 8, 3

4 Find the range of these sets of numbers.
   a 6, 0, 4, 4, 5, 8, 3, 2, 2, 6
   b 110, 150, 100, 120, 500

5 The ages of the members of an extended family are 2, 5, 16, 34, 36, 59, 65, 96.
   a Calculate the mean age.
   b Calculate the mean age in 10 years (assuming all are still living).

6 A class was given a maths test and the scores obtained were:
   Boys: 5, 45, 48, 60, 60, 64, 64, 67, 69, 79, 87, 88
   Girls: 40, 48, 54, 59, 60, 62, 64, 65, 70, 76, 84, 96
   a Calculate the mean score for **i** the boys and **ii** the girls.
   b Calculate the mean score for the whole class.

7 The mean age of four children is 7.5. Three of the children are aged 3, 4 and 8. Calculate the age of the fourth child.

Incorporating exercise:	1A	Key words	
Homework:	1.1	mean	median
Example:	1.1	measure of location	mode

Learning objective(s)

- use averages
- solve more complex problems using averages
- identify the advantages and disadvantages of each type of average and know which one to use in different situations

Prior knowledge

Pupils should know how to calculate the mean, mode, median and range from small sets of discrete data.

Starter

Encourage an initial discussion on averages. Ask pupils what is meant by an average. Explain that it is a single number used to represent all the data.

Ask pupils why we need (or use) three types of average.

Ask pupils to describe each of the three types of average.

Main teaching points

The first teaching point must be to go through finding the mode, mean and median of sets of discrete data. Look at data distributions with pupils where there is more than one mode to clarify how the mode is found in such situations. Similarly, it is also useful to clarify at this early stage how the median is found when the data consists of an even number of values.

Problems involving the mean are likely to focus on those such as questions 8–10 in Exercise 1A of the Pupil Book, where a mean is changed due to the addition of a number to the data, requiring this unknown number to be found. It is therefore suggested that you look at question 2 from worked examples 1.1 with the pupils.

The other aspect of this section is based on choosing the most appropriate type of average for the data. The mode and median tend to be more representative of the data when the data contains one or more outliers, as these are then ignored. For larger samples, any outliers tend to have less of an effect on the average, and the mean is more likely to be used in these cases. For qualitative data, the mode is the only measure of location that can be used. Remind pupils that measures of location – mean, mode and median – have units.

The mean is generally the most useful average to use since it is the only measure of location that uses *all* the data in its calculation. It is for this reason that the terms 'mean' and 'average' are often (though incorrectly) used interchangeably. The mean is also the foundation for much statistical work, a lot of which is beyond the scope of this book.

Plenary

Ask pupils to write down five numbers that have a range of 4, a mean of 4 and a median of 5. (Answer: 1 4 5 5 5. Others are possible, for example, 2 2 5 5 6.)

Incorporating exercise:	1B	**Key words**
Homework:	1.2	frequency table
Example:	1.2	

Learning objective(s)

- calculate the mode and median from a frequency table
- calculate the mean from a frequency table

Prior knowledge

Pupils should have covered Section 1.1 and be able to find the mode, mean and median of discrete data.

Starter

Ask pupils for definitions of each of the three types of average, in doing so recalling much of the material from the previous section.

Select a straightforward set of numbers with some repeated (such as 2, 2, 3, 3, 5), and ask pupils to calculate the mean. Do this with similar sets, leaving each set and each answer visible, then put each set into an appropriate tabular form, introducing pupils to the formality and notation of labelling the columns as x, f and fx.

Main teaching points

The finding of modes and medians from a frequency table is essentially revision. However, a majority of pupils often get confused in obtaining the median from a frequency table, and it would be useful to go through an example, perhaps question 1 from worked examples 1.2.

The new work in this section is in finding the mean from a frequency table. The standard method should be gone through with pupils in detail. Ensure pupils label the columns x, f and fx as required. When calculating the mean from the given values, encourage the constant use of the correct notation:

$$\bar{x} = \frac{\sum fx}{\sum f} = \cdots$$

Common mistakes

A common mistake at higher level is for pupils to use the formula:

$\bar{x} = \dfrac{\sum fx}{\sum x}$ rather than the correct $\bar{x} = \dfrac{\sum fx}{\sum f}$.

If columns are clearly labelled and the correct formula written down, this should not occur.

When using a calculator, there are two common mistakes. First, pupils often forget that if they make a mistake and start again, they must be very sure that they have cleared the calculator's memory before re-entering any data. Second, they must be sure which way round *their* calculator accepts the data value and frequency inputs. Again, calculators are often different in this respect.

Differentiation

Low achieving pupils will find this work difficult, or at least difficult to remember accurately. They may be more successful if the data is taken out of tabular form and listed in full before any calculations are attempted, although with a larger sample space this is likely to prove impractical.

Plenary

If pupils have access to scientific calculators, a useful plenary would be to demonstrate the method of finding a mean using the statistics mode of a calculator. The usefulness of this approach could easily be demonstrated by checking the answers of some of the questions in the Pupil Book or on the Worked examples sheet.

Incorporating exercise:	1C	**Key words**
Homework:	1.3	continuous data
Example:	1.3	discrete data groups
		estimated mean modal group

Learning objective(s)

- identify the modal group
- calculate and estimate the mean from a grouped table

Prior knowledge

Pupils should know how to calculate the mean from a grouped frequency table containing discrete data, as covered in Section 1.2. They should also be familiar with the use of inequality symbols.

Starter

It is often the case that revision will be needed on inequality symbols. Go through, $<$, $>$, $\leq$ and $\geq$ with the pupils.

Discuss a set of intervals given in terms of inequalities. For example, ask pupils what $0 \leq x < 10$, $10 \leq x < 15$, etc. mean. It might be useful to introduce and explain terms such as the lower class boundary and upper class boundary of each interval.

Explain why it is best to use inequalities when describing continuous intervals, rather than simply expressing a range such as 0–10, 10–15, etc.

Main teaching points

Pupils should not have any difficulty in finding the modal group from a grouped frequency table, as this is identical to finding the mode from a frequency table, which is covered in Section 1.2. A reminder that the modal group is found from the group with the highest corresponding frequency is usually sufficient.

The method of estimating the mean is the same as that in Section 1.2, with the additional step of taking x as the mid-interval value of each class. Formally, x is taken to be the mean of the lower and upper class boundary of each class interval.

Common mistakes

Some pupils decide to interpret the phrase "find an estimate for the mean", as meaning guess the mean, especially after a period of time following the work actually being taught.

Plenary

Discuss with pupils the reasons why class widths of a certain size might be chosen before collecting a sample. Must they all be of the same width? Under what circumstances might it make sense to choose different widths?

Examine the change in estimate of the mean if different class widths are chosen. For example, look at question 2 from Homework 1.3 but using these three class intervals: $0 \leq x < 4$, $4 \leq x < 8$ and $8 \leq x < 12$.

Incorporating exercise:	1D
Homework:	1.4
Example:	1.4

Key words

continuous data   histogram
discrete data
frequency
    polygon

Learning objective(s)

● draw frequency polygons for discrete and continuous data
● draw histograms for continuous data with equal intervals

Prior knowledge

Pupils should have covered Sections 1.1–1.3 of the Pupil Book. Although this is not essential in terms of drawing frequency polygons and histograms, this section is a natural continuation of the work done previously.

Starter

Statistical data is often presented in pictorial or graphical form. Ask pupils for the types of such presentations they may have already encountered.

Can they suggest any advantages one format might have over another? For example, a pie chart, which illustrates frequency as an area, could be argued to be more 'visually correct' than a bar chart, where the bar widths may be misleading to the eye.

Main teaching points

A **frequency polygon** is a graphical display of a frequency table. It is important that pupils are clear about where to plot the points comprising a frequency polygon. The height of each point is straightforward, as it just represents the frequency, but the horizontal position is more complex. For ungrouped data, the position is directly above the data value, whereas for grouped data, pupils should understand that they need to find the midpoint of each group as if they were going to calculate the mean, and then use this as the horizontal position to plot. Horizontal axes should always be drawn with a continuous scale, whether grouped or ungrouped data is being used.

Histograms give a visual representation of continuous data, and hence there are no gaps between bars (except in cases where the frequency is zero). The frequency of the data in a histogram is given by the area of a bar, as opposed to the case of the bar chart, where the frequency is indicated by the height of a bar. The y-axis is always labelled 'frequency density'. The frequency density for each class interval can be calculated from this formula:

$$\text{frequency density} = \frac{\text{frequency}}{\text{class width}}$$

The class width for each interval is found by subtracting the lower class boundary from the upper class boundary.

Common mistakes

The most common mistake in drawing frequency polygons is the use of grouped labels on data axes rather than continuous scales; for example, labelling a whole section of the axes '5 to 10' or '5–10' instead of a clear 5 at one mark and 10 at the next mark, with consistently sized gaps between the marks.

In drawing histograms, calculating the bar widths is straightforward if class intervals are described using inequalities, as the upper and lower class boundaries are given explicitly. However, if class intervals are not defined by inequalities, then care must be taken to calculate the bar widths correctly. For example, if the times of an event are noted to the nearest second, and the intervals are given as 6–10 secs, 11–15 secs, 16–20 secs, etc., then the first bar width should be 5 (10.5 – 5.5) and *not* 4 (10 – 6).

Although in this section all class intervals involving histograms will be of equal size – and, hence, bar widths will be of the same size – this shouldn't be assumed, as pupils are soon introduced to histogram problems involving unequal class intervals. Calculated bar widths must therefore be checked.

Differentiation

The principal differentiation here is with the type of data used. Drawing frequency polygons for ungrouped data is a grade D topic, whereas if the data is grouped it becomes a grade C topic. Drawing histograms of grouped data (with equal class intervals) is a grade C topic, but if the data has unequal class intervals (covered in Section 11.5) it is a grade A topic.

Plenary

Refer pupils to Worked example 1.3 with its grouped data on the height of seedlings. What coordinates would need to be plotted to obtain a frequency polygon from the data of the seedlings' heights?

Ask pupils to draw a frequency polygon and histogram of this data, asking them to compare and contrast the results.

Incorporating exercise:	1E
Homework:	1.5
Example:	1.5

Key words

class interval	lower quartile
interquartile	median
range	upper quartile

Learning objective(s)

● draw and read histograms where the bars are of unequal width
● find the median, quartiles and interquartile range from a histogram

Prior knowledge

Pupils should be familiar with the work covered in Section 1.3, namely estimating the mean from a grouped frequency table, and Section 1.4, drawing histograms from grouped data with equal class intervals.

Starter

Draw a normal distribution curve on the board. Explain that many real-life situations follow this distribution. For example, people's heights, weights of biscuits made by machines and so on, which is why it is called the normal distribution.

Ask pupils how they could find the median and quartiles. Establish that these are the values where the area under the curve is split into 25% sections.

Main teaching points

Now draw any histogram (with or without equal widths) on the board and ask how we could use the previous idea to find the median and quartiles.

Establish that the median is the value where the area on each side is half of the total area.

Discuss how to find this if the value is within the width range of a bar:

● Imagine the class boundaries are p, q, etc, and that the areas of the bars are A, B, etc.
● By working out the cumulative areas it can be established which bars the quartiles and medians are in.

Tell the class to imagine that the median is between t and u on this histogram:

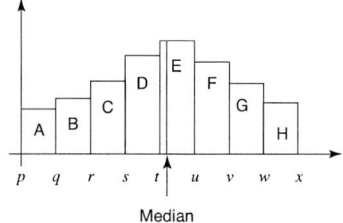

If a formula is needed this is:

$$\text{Median} = t + \left[\frac{\left[\dfrac{\text{Total area}}{2} \right] - (A + B + C + D)}{E} \right] \times (u - t)$$

Similar formulae are used for the lower and upper quartiles.

Plenary

Two ways of displaying continuous grouped data with unequal class intervals have now been considered, namely the frequency polygon and the histogram. Discuss with pupils which they think is the most useful. Or is the 'usefulness' dependent on the type of data being analysed?

Incorporating exercise:	1F
Homework:	1.6
Example:	1.6

Key words
moving average
seasonal trend
trend line

Learning objective(s)

● calculate a moving average and use it to predict future trends

Prior knowledge

No prior knowledge is required.

Starter

Give pupils a set of five numbers. After some explanation, ask them to calculate a set of two-point moving averages. Can they suggest a reason how or why this procedure might be useful?

Main teaching points

A moving average gives the general trend for a set of data over time. For example, the sales of ice creams can fluctuate greatly over the course of one year, but year on year there may be a general trend of increased sales.

General trends in data that might be subject to short-term fluctuations can be seen more clearly by calculating moving averages, which have the effect of smoothing out these monthly (or seasonal) fluctuations. Stockbrokers, for example, often use six-monthly moving averages as a guide to the general trend of a company's share price.

Given a sequence of numbers a_i, $(i = 1, \ldots, n)$, the sequence of two-point moving averages would be defined as:

$$\frac{1}{2}\left(a_1 + a_2, a_2 + a_3, \ldots, a_{n-1} + a_n\right)$$

The sequence of three-point moving averages would be:

$$\frac{1}{3}\left(a_1 + a_2 + a_3, a_2 + a_3 + a_4, \ldots, a_{n-2} + a_{n-1} + a_n\right)$$

Similarly, you can calculate the four-point, five-point and six-point moving averages, and so on.
The trend line is found by plotting the moving averages against the time and joining the points with straight lines.

Plenary

As mentioned in the Main teaching points, ice cream sales inevitably increase every summer and decrease in the winter. The general trend in year-on-year sales can be analysed through moving averages. What other products might benefit from a moving average analysis? (There are many possibilities, such as flowers, toys, Easter eggs, etc.)

Incorporating exercise:	1G, 1H	Key words
Homework:	1.7	data collection leading question
Example:	1.7	sheet survey
		hypothesis

Learning objective(s)

- conduct surveys
- ask good questions in order to collect reliable and valid data

Prior knowledge

Pupils must be able to construct and use tables. Pupils should know how to use a tallying procedure accurately and effectively.

Starter

Ask pupils to count up how much money they have on them, but not to tell you. Give them a (deliberately poor) choice of possible responses, such as **(i)** between £0 and £1, **(ii)** between £1 and £5 or **(iii)** between £5 and £10. Collect up their responses by tallying on the board.

Ask for comments. Points to look for include **(a)** why not just ask exactly how much everybody has, **(b)** what do you do if you have (for example) exactly £5, and **(c)** what do you do if you have more than £10?

Main teaching points

Pupils should understand that simple, clear and unambiguous questions, with a clear choice of the possible answers, are the most appropriate and efficient way to structure a data collection sheet. The design should include appropriate space to record the responses. This will usually be spaces for summarising responses by tallying.

Pupils also need to consider the most appropriate way to design questions and response options when designing a questionnaire. They should understand the five main rules that are highlighted in the Pupil Book. Plenty of discussion is recommended here, so that pupils have every opportunity to consider what makes good and bad designs. Pupils should be given the opportunity to raise subjects of their own choosing for the questionnaire design, as they are more likely to have a clear understanding of the salient points. However, their interest may also give them preconceived ideas about how they hope people would respond. As a result, they may well include leading questions, and these should be clearly pointed out and addressed in discussion.

Common mistakes

The most common mistake here is in the design of response options. In particular, the options to respond at either extreme are often forgotten or not well handled. Lower achieving pupils also often find it difficult to find faults with data collection sheets or survey questions that they are presented with.

Plenary

Ask for suggested response options for these topics (or others of your own choice):
- how often people use the Internet
- how many days off work/school people have had in the last year.

Look for a sensible range of response options, with no overlaps and the extremes catered for realistically.

Incorporating exercise:	1I
Homework:	1.8

Key words

margin of error	Retail Price Index
National Census	social statistics
Polls	time series

Learning objective(s)

● be introduced to some of the more common social statistics in daily use

Prior knowledge

Pupils need to be able to interpret information from dual bar charts.

Starter

Ask pupils if they know how many people live in Britain. The answer is approximately 60 million.

Discuss how we know this. What is the exact figure? Can the exact figure ever be found?

Mention the National Census and discuss what sort of data this collects (number of people per house, ages, gender, religion, ethnicity, education, and so on), how often it is taken (every 10 years) and if it is compulsory (yes).

Main teaching points

Most statistics that are met on a daily basis in newspapers are 'social'. For example, percentages of unemployed, increases in the cost of living, changes in prices over time, populations and spending on the NHS.

Discuss each of these and make sure that pupils understand the idea of an index and a base number. Why is it usually 100? Review percentages if necessary, in particular the percentage multiplier.

Talk about time series. In particular, emphasise that the lines joining points represent trends and not actual values, which are generally measured at a specific time each year or month. Examples would be temperature changes, cost of houses and exchange rates.

Differentiation

This is a new topic for many and may be a difficult concept for some pupils. If necessary discuss the questions in the exercise to give pupils some idea of how to do them.

Plenary

Sketch the following graph on the board, which shows a rough distribution of ages in Britain in 1950 and 2000.

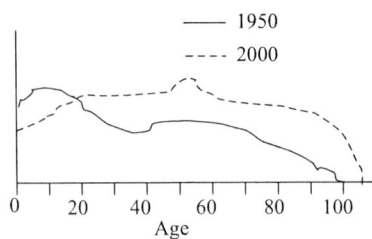

Ask pupils what use such information might be to a government? Explain that the blip is the post war baby boom.

Discuss how an aging population means pension problems and fewer young people means less school places will be needed in 5 years time.

1.9 Sampling

Incorporating exercise:	1J
Homework:	1.9
Example:	1.9

Key words

population   stratified
random   unbiased
sample

Learning objective(s)

- understand different methods of sampling
- collect unbiased reliable data

Prior knowledge

There is no direct prerequisite for studying sampling methods, although clearly this section naturally complements Section 1.7 on surveys.

Starter

Say to the class you are going to do a quick survey on their television-watching habits. Say you will choose five people totally at random. Proceed to pick five girls (or boys) and ask fatuous questions on the previous night's programmes. At the end, ask the open question, "What was wrong with my sample?" Pupils may argue it was not completely random, but you can argue otherwise! Ask the class how a more appropriate sample might be chosen. Must the sample be exactly half boys and half girls?

In a single-sex class, you could make a point of questioning the five eldest pupils in the class. Age as well as sex is likely to play a part in determining viewing habits, especially if the sample contains a significant age range.

Main teaching points

There are two main types of sampling to be considered: random sampling and stratified sampling.

In a random sample, a sample is chosen on the basis that all members of the population are equally likely to be chosen.

In a stratified sample, the sample is divided into subgroups, or strata. The size of each strata is chosen in such a way that it is in the same proportion as it is in the actual population.

You would choose stratified sampling over pure random sampling if you have some evidence (or a strong intuition) that subgroups may have significant differences in the characteristic you are trying to measure. For example, it is known from an analysis of previous surveys that there is significant difference between the voting intentions of men and women. So if a sample is to be taken to examine people's voting intentions in the next general election, it makes sense to stratify the sample by sex. Suppose you know that the population as a whole (the electorate in this case) consists, say, of 56% females and 44% males. If the chosen sample is to be 50 people, then you should ensure that 28 of them are female and 22 male. These in turn would be selected randomly.

Plenary

Two other methods of sampling are systematic sampling and attribute sampling. Ask pupils to investigate both of these and suggest possible advantages and disadvantages of each method.

Overview

2.1 Line graphs
2.2 Stem-and-leaf diagrams
2.3 Scatter diagrams
2.4 Cumulative frequency diagrams
2.5 Box plots
2.6 Measures of dispersion

This chapter covers the remaining statistical topics required at this level. The first three sections are also covered in the Foundation syllabus and are identical in content. At higher level, the essential section(s) are on cumulative frequency diagrams (and their interpretations) and corresponding box plots. The final section on measures of dispersion is essentially based upon the standard deviation. This is no longer examined at GCSE, although since it is a more useful measure of dispersion than the interquartile range, it can be used in the Data handling task of the coursework.

Context

The normal distribution used to be covered at GCSE, and it may be advantageous to go over this briefly with the more able pupils. It would provide a relevant example of how the mean and standard deviation can relate to real life, such as the modeling of the distributions of heights, test scores, or the ages of members of a population. The expectation is that 66% of the population lie within one standard deviation of the mean and that the range is covered by approximately six standard deviations.

AQA B references

AO4 Handling data: Processing and representing data

2.1–2.5   4.4a "draw and produce, using paper and ICT, pie charts for categorical data, and diagrams for continuous data, including line graphs (time series), scatter graphs, frequency diagrams, stem-and-leaf diagrams, cumulative frequency tables and diagrams, box plots and histograms for grouped continuous data"

Route mapping

Exercise	D	C	B	A	A*
A	all				
B	all				
C	1	2–5			
D			all		
E			1–5	6	
F					
G					

Answers to diagnostic Check-in test

1 $(-2.5, -1.5)$

2 a Gradient is positive ($= 1$)   **b** Gradient is negative ($= -1$)   **c** Gradient is 0   **d** Gradient is infinite

3 a $20 \leq L < 30$ **b** $31.7\,\text{cm}$ **c** $30 \leq L < 35$

1 Draw a set of *x-y* axes, marking each axis from –4 to +4. The coordinates (3.5, 1.5), (3.5, –1.5) and (–2.5, 1.5) are three points of a rectangle. Plot them and state the coordinates of the missing point.

2 What can you say about the gradients of each of these lines?

   a $y = x$

   b $y = -x$

   c $y = 3$

   d $x = 3$

3 The table shows the lengths of some marrows grown on Lucy's allotment.

Length (*L* cm)	Frequency (*f*)
$0 \leq L < 20$	4
$20 \leq L < 30$	16
$30 \leq L < 35$	12
$35 \leq L < 40$	10
$40 \leq L < 50$	6
$50 \leq L < 60$	2

   a State the modal group.

   b Estimate the mean length of a marrow.

   c In which class interval must the median lie?

Incorporating exercise:	2A	**Key words**
Homework:	2.1	line graphs
Example:	2.1	trends

Learning objective(s)

● draw a line graph to show trends in data

Prior knowledge

Pupils must be able to understand data presented in tabular form. Pupils need to be able to plan and read scales on axes in a variety of situations.

Starter

Present the class with a number of line segments, with different numbers of sections along them (two, four and five sections are the most usual). Put a number at each end of the line, and ask for the numbers that go at other positions along the line segment.

Repeat the exercise with times along the line segments. (A variety of time units is useful, such as minutes, years, dates, etc.)

Main teaching points

Pupils should understand that the use of line graphs to represent statistical information is usually restricted to time series. The reason for this is that line graphs allow estimates to be made at intermediate points.

Some discussion is useful on choosing suitable scales for the axes. If times are measured from a 'zero' or 'starting time', then the horizontal axis may start at zero; if times are years, or days of the week, or times of day, then it is impracticable to start at zero. Time is always plotted on the horizontal axis.

Pupils should understand that line graphs are useful because they give a picture of how something is changing over a period of time. This is what allows intermediate values to be estimated, and it may allow future predictions to be made, although these should always be treated very cautiously.

Common mistakes

The most common mistakes here are the poor choice, or incorrect use, of scales to represent times. It is also common for pupils to make intermediate or future estimates without giving realistic thought to the data.

Plenary

Have a classroom discussion about the value of line graphs, by asking pupils when they can be used, why they can be useful, etc.

Incorporating exercise:	2B
Homework:	2.2
Example:	2.2

Key words

discrete data	raw data
ordered	unordered

Learning objective(s)

● draw and read information from an ordered stem-and-leaf diagram

Prior knowledge

Pupils must be able to order a set of numbers. They must have a clear understanding of place value, in order to classify the digits of a set of numbers into the same category.

Starter

Show the class a set of numbers, and ask them to arrange them in order, lowest to highest. Do this for sets of numbers of different sizes. For example, you might choose a set of all two-digit integers, all two-digit decimal numbers, all numbers with one unit digit and one decimal digit, etc.

Show the class this set of numbers; 23, 54, 20, 48, 26, 58, 41. Ask them to suggest ways of selecting groups of numbers from this collection that have something in common. Examples include 54 and 48 (both are multiples of 6); 23 and 41 (both are odd numbers), etc. Hopefully, someone will suggest groups with the same tens digit (20, 23 and 26; 41 and 48; and 54 and 58).

Main teaching points

Pupils should understand that it is helpful to order the original data before constructing a stem-and-leaf diagram. It should also be made clear to them that the benefit of this method of representation is that it gives a clear indication of how the data is distributed. It shows the range of the data in a similar way to a grouped frequency table, but maintains all the detail of the original data.

Perhaps the most difficult skill here is in reading the data from a stem-and-leaf diagram. It would be beneficial for pupils to have as much practice in this as possible, ideally as a whole class activity so that it generates discussion and brings out the most common mistakes.

Pupils should understand that a stem-and-leaf diagram is much easier to read if the 'leaf' digits are well spaced, and kept in clear columns.

It should be noted that this technique will only be used with two significant figure numbers, but that the numbers may not be two-digit integer numbers. Pupils may have to cope with numbers such as 3.7, or 370, or even conceivably 0.37. Particular stress should be placed on the importance of the key in this technique.

Common mistakes

The most common mistake is in reading stem-and-leaf diagrams: pupils will often forget to recombine the two digits of the original data value, or will not do so consistently. They may also forget to use the key, and so make mistakes with the size of the numbers.

When creating a stem-and-leaf diagram, pupils often forget to provide a key. It is also common not to be careful enough with the spacing of the digits, reducing the effectiveness of the diagram.

Differentiation

Higher achieving pupils will find this section straightforward, and may benefit from being provided with more challenging collections of numbers, such as numbers in the hundreds or thousands, or even decimal numbers.

Plenary

Ask pupils to discuss the possible advantages or disadvantages of stem-and-leaf diagrams over grouped frequency tables.

Ask pupils to bring in one of their own CDs. Ask them to present the time durations of the tracks on their CD in the form of a stem-and-leaf diagram.

© HarperCollins*Publishers* Ltd 2006 Module 1: Data handling

Incorporating exercise:	2C
Homework:	2.3
Example:	2.3

Key words

line of best fit	positive
negative	correlation
correlation	scatter diagram
no correlation	variable

Learning objective(s)

- draw, interpret and use scatter diagrams

Prior knowledge

In terms of statistical work, there is no prerequisite for studying this section. Pupils should, of course, be able to use appropriate scales on axes and be able to interpret intermediate positions within the scales.

Starter

Draw a straight-line graph on a coordinate grid. Ask pupils to find the value of one variable that corresponds with your choice of the other. This might be best done with the help of an OHP.

Main teaching points

Pupils should be shown how to construct a scatter diagram and draw a line of best fit. Two corresponding data values are required for each point to be plotted, and these two values are used as one pair of coordinates.

The line of best fit should be representative of the underlying trend of all the data points. It should be a single straight line, and have approximately the same number of points on either side of the line. The sum of the distances from the points on one side of the line should be approximately the same as the distances to the points on the other side. Outliers should be ignored when drawing a line of best fit.

Pupils should understand how to interpret and use a scatter diagram. The concept of correlation should be discussed at some length. The aim should be to get pupils to consider whether two variables have any connection or influence on each other and, if so, whether the correlation is positive, negative or whether there is no correlation at all.

Pupils should understand how to use the line of best fit to find an approximate corresponding value.

Common mistakes

Pupils often draw lines of best fit inappropriately. Remind them that the line does not necessarily have to go through the origin, nor does it actually have to go through any plotted points.

Differentiation

Higher achieving pupils should be directed to focus on questions 2–5 of Exercise 2C of the Pupil Book, as these are the grade C questions. These require the greatest amount of interpretation of the scatter diagrams and the greatest understanding of the underlying concepts.

Plenary

Ask pupils for examples of variables which would give the different types of correlation. Try to get examples showing, respectively, positive, negative and no correlation.

	Incorporating exercise:	2D
	Homework:	2.4
	Example:	2.4

Key words

cumulative	lower quartile
frequency	median
diagram	upper quartile
dispersion	
interquartile	
range	

Learning objective(s)

⦿ find a measure of dispersion (the interquartile range) and a measure of location (the median) using a graph

Prior knowledge

Pupils should be familiar with estimating the mean from grouped continuous data. In particular, the distinction between lower and upper class boundaries and mid-interval values should be known.

Starter

Tell pupils that this section returns to grouped continuous data, so start by revising some terminology from Chapter 1 of the Pupil Book. Write some class intervals on the board, such as $0 \leq h < 10$, $10 \leq h < 20$ and ask pupils for the lower class boundary, the interval midpoint and the upper class boundary.

Go through a quick example of estimating the mean from a grouped distribution. Emphasise that estimating the mean from grouped data requires the plotting of mid-interval values, whereas in contrast estimating the median for that data involves using the upper class boundaries.

Main teaching points

A cumulative frequency curve (or polygon) allows the median to be estimated from grouped continuous data.

Cumulative frequencies (on the y-axis) should be plotted against upper class boundaries of class intervals (on the x-axis). If these points are joined with a smooth curve, the result is a cumulative frequency curve. If they are joined by a sequence of straight lines, the result is a cumulative frequency polygon.

Suppose total frequency is n. Then the lower quartile is the $\frac{n}{4}$ th value, the median is the $\frac{n}{4}$ th value and the

upper quartile is the $\frac{3n}{4}$ th value. These values may be found from reading across from the y-axis and finding

the corresponding x-value.

The interquartile range is the only measure of dispersion to be examined at GCSE other than the range itself. It is found by subtracting the lower quartile from the upper quartile. Its value indicates the spread of the middle 50 per cent of the data. A small interquartile range indicates the data is generally consistent about the median, while a larger value indicates the data is more spread out.

As a measure of dispersion, the interquartile range is advantageous as it eliminates extreme values through only considering the middle 50 per cent of data. However, this can also be seen to be a limitation and the interquartile range provides little basis for more advanced work in statistics.

Common mistakes

Lower achieving pupils will simply write down the median as $\frac{n}{2}$ rather than the $\frac{n}{2}$ th value: in other words, they will not make use of the cumulative frequency curve. Some pupils will also plot class mid-intervals along the x-axis rather than the correct upper class boundaries, confusing their work with the estimating of the mean.

Differentiation

The topic of cumulative frequency curves is generally grade B standard at GCSE. However, the procedure for completing a table, drawing the curve and finding the median and interquartile range for a set of data is not difficult and should be manageable by the vast majority of pupils, even if it is rote-learned by the less able.

A standard discriminator used by examiners is to ask for an estimate of numbers of data greater than a certain value. An example of this type of question is provided in the Worked examples.

Plenary

Which is better, a high interquartile range or a low interquartile range? Ask pupils for their opinions. Can they think of practical instances where it would be preferable to obtain:
- a low interquartile range?
- a high interquartile range?

Incorporating exercise:	2E
Homework:	2.5
Example:	2.5

Key words

box plot	lowest value
highest value	median
lower quartile	upper quartile

Learning objective(s)

● draw and read box plots

Prior knowledge

Pupils should have covered Section 2.4 in the Pupil Book and be able to find the lower quartile, median and upper quartile of a set of continuous grouped data from a cumulative frequency diagram.

Starter

Ask pupils to recall what can be found from a cumulative frequency curve (that is, the lower quartile, median and upper quartile). Can they suggest a precise definition as to what the lower quartile actually is or means? In a similar way, try to encourage suggestions of how the median and upper quartile might be defined.

Main teaching points

Box plots, or box-and-whisker plots, give a visual representation of the range of a set of data and the location of the middle 50 per cent of the data.

The left-hand side of the box corresponds to the lower quartile and the right-hand side to the upper quartile. The median is indicated within the box by a vertical line.

The left-hand line or whisker extends down to the lowest value taken by the data and the right-hand line or whisker extends to the greatest value taken by the data. If the data is grouped, for example, using the intervals $0 \leq h < 10, \ldots, 40 \leq h < 50$, then the left-hand line extends down to 0 and the right-hand line extends up to 50.

The three quartiles Q_1, Q_2 and Q_3 (lower quartile, median and upper quartile) can also be used to find the skewness of a set of data. (Although skewness is not explicitly part of the syllabus, it is referred to in question 5, Exercise 2E of the Pupil Book.)

Essentially:
● if $Q_2 - Q_1 < Q_3 - Q_2$ then the data has a positive skew
● If $Q_2 - Q_1 = Q_3 - Q_2$ then the data is symmetric
● If $Q_2 - Q_1 > Q_3 - Q_2$ then the data has a negative skew.

Common mistakes

If class intervals are described using inequalities, then pupils sometimes take the least and greatest values – the ends of the whiskers – to be the midpoints of the first and last class intervals respectively.

Plenary

Discuss with pupils the advantages or disadvantages a box plot might have compared with a cumulative frequency diagram.

Incorporating exercises:	2F, 2G	Key words	
Homework:	2.6	dispersion	ungrouped data
Example:	2.6	measure of spread	
		standard	
		deviation	

Learning objective(s)

◉ calculate standard deviation for a set of data

Prior knowledge

Pupils should be familiar with the process of estimating the mean from grouped continuous data. They should also be able to find an estimate for the interquartile range of grouped data from a cumulative frequency diagram and appreciate its use when comparing two sets of data.

Starter

Two measures of dispersion have already been encountered: the range and the interquartile range. Ask pupils to define carefully what is meant by each. What are the advantages of the interquartile range as a measure of dispersion compared to the range? Does it have any disadvantages?

Main teaching points

The standard deviation gives a measure of dispersion of the data about the mean. A direct analogy would be that of the interquartile range, which gives a measure of dispersion about the median. However, the standard deviation is generally more useful than the interquartile range as it includes *all* data in its calculation. In contrast, using the interquartile range immediately discounts 50 per cent of the data.
The standard deviation σ of a set of data is defined as:

$$\sigma = \sqrt{\frac{\sum f(x-\bar{x})^2}{\sum f}}$$, where $\bar{x}$ is the mean and f the frequency

An equivalent definition is:

$$\sigma = \sqrt{\frac{\sum fx^2}{\sum f} - \bar{x}^2}$$

Either formula can be used when calculating the standard deviation using a tabular method or frequency table.

Plenary

Instead of using a calculator, find the standard deviation of Lucy's marrows (question 2 of Homework 2.6) by using the formula:

$$\sigma = \sqrt{\frac{\sum f(x-\bar{x})^2}{\sum f}}$$

The best way to do this is to add and complete two extra columns to the frequency table for $(x-\bar{x})^2$ and $f(x-\bar{x})^2$ and then compute the total of the latter column. Verify that the answer is 10.1 cm and thus show that both methods are correct. Note, however, that the calculator method is much quicker and more efficient.

What other advantages does the calculator method have over the tabular method? (There is no loss of accuracy and the estimated mean does not need to be calculated first.)

Probability

Overview

This chapter covers all the probability material in the Higher syllabus. From use of the language associated with probability, and the ideas of experimental probability and expectation, it develops through to the ideas of mutually exclusive events, independent events and probabilities associated with combined events. Tree diagrams are used to calculate probabilities of multiple independent events, before finally looking at problems relating to dependent events.

Context

The subject of probability is one which pupils can readily link to their own experiences in life. With the recent increase in games of chance, such as the National Lottery, pupils are likely to have experiences which they can relate to the theory covered in this chapter.

It is helpful throughout the study of probability to use the language of chance and prediction, so that pupils are reminded that probabilities do not predict results in an exact way, only the likelihood of particular results.

AQA B references

AO4 Handling data: Interpreting and discussing results

3.1   4.5i "understand that if they repeat an experiment, they may – and usually will – get different outcomes, and that increasing sample size generally leads to better estimates of probability and population parameters"

AO4 Handling data: Processing and representing data

3.2   4.4d "identify different mutually exclusive outcomes and know that the sum of the probabilities of all these outcomes is 1"

3.3   4.4b "understand and use estimates or measures of probability from theoretical models, or from relative frequency"

3.4   4.4c "list all outcomes for single events, and for two successive events, in a systematic way"

3.5, 3.6, 3.8   4.4g "know when to add or multiply two probabilities: if A and B are mutually exclusive, then the probability of A or B occurring is $P(A) + P(B)$, whereas if A and B are independent events, the probability of A and B occurring is $P(A) \times P(B)$"

3.7–3.9   4.4h "use tree diagrams to represent outcomes of compound events, recognising when events are independent"

Exercise	F	E	D	C	B	A	A*
A				all			
B				1–7	8–9		
C				1–8	9		
D			1–5	6–7			
E		1	2–5	6–10			
F			1–5	6–9			
G					all		
H					1–6	7	
I						all	
J						all	
K						1–2	3–11

Answers to diagnostic Check-in test

1 a 30% **b** 77% **c** 94% **d** 9%

2 a $\frac{1}{3}$ **b** $\frac{5}{8}$ **c** 0.6 **d** 0.82

e $\frac{7}{12}$

3 a $\frac{3}{4}$ **b** $\frac{7}{10}$ **c** $\frac{5}{6}$ **d** 0.91

e 0.68 **f** 0.39

4 a $\frac{18}{125}$ **b** 20% **c** $\frac{8}{25}$ **d** 500

5 a 24 **b** 50 **c** 135 **d** 112

1 For each of these percentages, work out how much must be added to reach 100%.

   a 70%   **b** 23%   **c** 6%   **d** 91%

2 For each fraction or decimal, work out how much must be added to reach 1.

   a $\dfrac{2}{3}$ **b** $\dfrac{3}{8}$ **c** 0.4

   d 0.18   **e** $\dfrac{5}{12}$

3 Work out each of these, simplifying your fraction answers wherever possible.

   a $\dfrac{1}{8} + \dfrac{5}{8} =$

   b $\dfrac{3}{10} + \dfrac{2}{5} =$

   c $\dfrac{2}{3} + \dfrac{1}{6} =$

   d $0.32 + 0.59 =$

   e $0.28 + 0.4 =$

   f $0.3 + 0.09 =$

4 The travellers on a ferry were asked the reason for their journey. The results were put into this table.

Reason for journey	Holiday	Business	Shopping	Family visit	Other
Frequency	40	25	20	18	22

   a What fraction of the travellers were visiting family?

   b What percentage were on business?

   c What fraction were on holiday?

   d If 2500 people use the ferry in a month, how many, based on these results, would you expect to be travelling on business?

5 Work out each of these calculations.

   a $\dfrac{1}{4}$ of 96

   b $\dfrac{2}{3}$ of 75

   c $\dfrac{3}{5}$ of 225

   d $\dfrac{7}{8}$ of 128

			Key words	
Incorporating exercise:	3A		experimental	trials
Homework:	3.1		probability	
Example:	3.1		relative	
			frequency	

Learning objective(s)

- calculate experimental probabilities and relative frequencies
- estimate probabilities from experiments
- use different methods to estimate probabilities

Prior knowledge

Pupils must know how to find the probability of an event, using the definition $P(A) = \dfrac{\text{Number of ways } A \text{ can happen}}{\text{Number of possible outcomes}}$, where A is an event.

Starter

Prepare a bag or box containing a number of different coloured balls or counters, or similar. Ask pupils to select one without looking and put it back. Have a pupil collect the data as it is generated. At different stages, for example after four selections and again after 10, ask for suggestions about what is in the bag. Ensure that pupils give reasons for their suggestions.

Main teaching points

Pupils should understand that there are three different methods for finding probabilities and they should know which method to use in any given situation. The three methods are:
- equally likely outcomes
- experiment or survey
- use of historical data.

Pupils should understand that experimental probability increases in accuracy with increasing numbers of trials.

It is worth at this stage reminding pupils that probabilities may be expressed as fractions or decimals or percentages. While fractions may be the most convenient way to express the probabilities in small practical exercises, it is often more useful to use one of the other representations for experimental probabilities as they are easier to compare as the number of trials in the experiment increases.

Common mistakes

When data is presented in tabular form, pupils often have difficulty in identifying the correct information. It should be made clear to them that they need to use 'number of correct results' and 'number of trials' in all cases, but they need to be able to interpret a table correctly to find these values. Some tables will be cumulative frequencies of the required outcome and others will be frequencies of all the results.

Plenary

Ask pupils to feed back the results of question 6, Exercise 3A in the Pupil Book. It would be hoped that by collating results, it can be shown that more accurate experimental probabilities may be obtained.

Incorporating exercise: 3B	**Key words**
Homework: 3.2	complementary
Example: 3.2	exhaustive
	mutually
	exclusive

Learning objective(s)

● recognise mutually exclusive, complementary and exhaustive events

Prior knowledge

It would not be a bad idea to remind pupils of the make up of a deck of cards. While this may have been common knowledge at one time, it might be a little rash to assume all pupils will be this familiar with them today. There is also the question of jokers in the pack. Unless stated otherwise, pupils should assume a pack consists of 52 cards, with no jokers.

Starter

Continuing in the same vein as the Prior knowledge ask pupils simple probability questions based around a pack of cards. For example, suppose a single card is taken from a pack. What is the probability that:

  a it is a 2 of clubs?   **d** it is a royal card?
  b it is a spade?   **e** it is an ace?
  c it is a red card?   **f** it is an odd-numbered card?

Ask for all probabilities to be given as simplified fractions.

Main teaching points

Two events A and B are said to be *mutually exclusive* if they cannot happen at the same time.

In particular, if B is the event of 'A *not* happening', then we say that A and B are *complementary events*, or that B is the *complement* of A. In this case we have

$P(B) = 1 - P(A)$, or $P(A) + P(B) = 1$.

Since in this example, there is no possibility of anything other than A or B happening, events A and B are said to be *exhaustive*.

The majority of questions in Exercise 19B in the Pupil Book relate to the understanding of two events A and B being mutually exclusive. Question 5, however, requires possible outcomes to be listed and the probabilities computed using the formula:

$$P(A) = \frac{\text{Number of ways } A \text{ can happen}}{\text{Number of possible outcomes}} \text{ for a given event } A$$

Common mistakes

Mistakes or misunderstandings inevitably occur within the vicinity of fractions. For questions involving mutual exclusivity, ensure pupils are familiar in being able to do calculations requiring the application of $P(B) = 1 - P(A)$. Run through a few examples, getting pupils to evaluate forms such as:

$1 - \dfrac{7}{12}$ and $1 - \dfrac{13}{25}$

Plenary

Return to the pack of cards. Suggest to pupils that a single card is picked at random. Ask them for descriptions of two events which are:

a mutually exclusive and exhaustive
b mutually exclusive and not exhaustive
c not mutually exclusive.

Incorporating exercise:	3C	
Homework:	3.3	
Example:	3.3	

Key words
expectation

Learning objective(s)

● predict the likely number of successful events given the number of trials and the probability of any one event

Prior knowledge

Pupils must know how to find the probability of an event. They must also know how to calculate a fraction of a quantity.

Starter

Recap, or ask a pupil to explain, how to find a fraction of a quantity.

Display the number 240 and ask pupils to choose a fraction to find of this number. It may be more appropriate to limit the choice of fractions to be used to halves, thirds, quarters, fifths, sixths, eighths, tenths, twelfths.

Main teaching points

Pupils should understand that probabilities are a measure of the proportion of trials that are expected to conclude in the outcome under consideration. It is important that they understand that reality rarely follows exactly what is expected in theory, but that as the number of trials increases, the proportion should become closer to the expected or predicted proportion.

It is worth putting this in context and giving pupils examples of where this theory is used in daily practice. The National Lottery, for example, relies entirely on expectations. Also, games of chance, such as fairground games, would not operate if the owners could not be confident of making a profit out of them in the long run. The calculation of expected profit from any game of chance relies entirely on a large number of participants ensuring that the proportion of winners to losers is very close to what the theoretical probability suggests.

Common mistakes

Mistakes here are likely to be confined to one of two kinds. Firstly, some pupils work out the required probability incorrectly; secondly, some work out the fraction of the number of trials incorrectly.

Differentiation

Question 8 of Exercise 3C in the Pupil Book will stretch all but the most able pupils, as it requires a little more creativity of thought than just following a prescribed method to find an answer.

Plenary

Display a sample space diagram of total scores from the roll of two dice. Ask the pupils how many times they would expect to get different scores (such as 4, 12, 5, less than 4, etc.), from 360 rolls of the dice. Which score would they expect to get most often? Ask them why you picked 360 rolls.

Incorporating exercise: 3D

Homework: 3.4

Example: 3.4

Key words
two-way tables

Learning objective(s)

● read a two-way table and use them to work out probabilities and interpret data

Prior knowledge

Pupils must be able to retrieve information presented in tabular form. They must know how to find the probability of an event. They must be able to find the expected number of results from the total number of trials and the probability of the event in question.

Starter

Collect some data from the class that can be tallied into a two-way table. Ask pupils, say, to name their favourite pop group, or football team, or sport, from a given short list. In this way, you can construct a two-way table with column headings male and female, and, say, four row headings. (In a single-sex class, find some other way of splitting the class into two groups.)

Ask the class to provide probabilities of selecting different classes of people (for example, a girl who supports Man Utd, or a boy whose favourite sport is golf, etc). Choose a relatively straightforward multiple of the number of pupils in the class, and ask how many would be expected out of this total to be in each category in the table.

Main teaching points

Pupils should understand that a two-way table is a means of classifying or categorising a population into subsections. They will need to understand that the size of the population can be found simply by adding all the entries in the table together. The probabilities associated with each category in the table are the fractions that each category represents of the total population.

Pupils should understand that these probabilities can be used to find the expected numbers of other populations, provided that the two populations represent similar characteristics.

Common mistakes

Most mistakes here will come from misunderstandings in how to read the data from a two-way table. Going through many examples, with as much discussion as possible, will assist pupils' understanding of how to interpret these tables.

Differentiation

The questions in Exercise 3D in the Pupil Book cover grades C and D. More able pupils should find question 1 easy and questions 2 to 4 also quite easy. Less able pupils will find questions 5 and 6 difficult and question 7 very difficult.

Plenary

Revisit the information from the lesson starter. Ask pupils to suggest how to calculate the percentages of the total that each represents.

Incorporating exercise:	3E
Homework:	3.5
Example:	3.5

Key words
either

Learning objective(s)

● work out the probability of two events such as $P(\text{event } A)$ or $P(\text{event } B)$

Prior knowledge

Pupils must know how to find the probability of an event. They must also know how to express one number as a fraction of another, how to simplify fractions and how to add and subtract fractions.

Starter

Display the fractions $\frac{1}{2}, \frac{1}{3}, \frac{1}{4}, \frac{1}{6}$ and $\frac{1}{12}$. Ask pupils to choose a pair to add. Then ask them to choose any three to add, then any four to add, and finally to add all five.

Main teaching points

Pupils should understand that the probability of either of two (or more) mutually exclusive outcomes can be found by adding together the probabilities of each of the individual outcomes. So, if events A and B are mutually exclusive, then $P(A \text{ or } B) = P(A) + P(B)$.

The concept of mutual exclusivity should be revisited from Section 3.2, along with examples of outcomes which are and are not mutually exclusive to illustrate the difference. They should also understand why it is not appropriate to add together probabilities of outcomes which are not mutually exclusive. It may be useful to illustrate this with a suitable example. For example, if a dice is rolled, consider the probability of rolling an even number or a prime number.

Pupils should understand why probabilities of 'either one outcome *or* another outcome' are bigger than each individual probability, and why probabilities of 'one outcome *and* then another outcome' are smaller than each individual probability.

Common mistakes

Pupils often make mistakes in adding together probability fractions.

Pupils often learn how to add together probabilities, but they do not always understand when it is appropriate to do so. Question 8 of Exercise 3E is particularly useful in illustrating an occasion when it is not appropriate to do so.

Plenary

Display a target board with whole numbers on it. Ask for probabilities of events such as 'picking a multiple of 4 or 6', 'picking a prime number or a factor of 100', 'picking a number which is a factor of 40 and 60', etc. The choice of questions will depend on the numbers displayed.

Incorporating exercise:	3F	**Key words**
Homework:	3.6	probability
Example:	3.6	space diagram
		sample space
		diagram

Learning objective(s)

● work out the probability of two events occurring at the same time

Prior knowledge

Pupils must know how to find the probability of an event.

Starter

Split the class into pairs. Provide each pair with two dice. With a time limit of 2 minutes, ask each pair to roll the dice, add the scores on the dice on each roll, and record each total score. While they are doing this, prepare a tally chart on the board, and then collate all the results from the class when they have finished. Ask for the relative frequencies, or experimental probabilities, for each score.

Main teaching points

Pupils should first clearly understand what is meant by combined events.

They should understand how to construct a sample space diagram for combined events. This is a method for easily identifying all the possible equally likely outcomes. From this knowledge, the method for finding a theoretical probability is now simply:

$$P(A) = \frac{\text{Number of ways } A \text{ can happen}}{\text{Number of possible outcomes}}, \text{ where } A \text{ is an event}$$

Pupils should understand that all the other techniques learned in this chapter can be applied equally to combined events as well, including the addition rule, experimental probability, and the probability that an event will not happen.

Common mistakes

The most common mistake with this type of problem is that pupils are careless about reading the instructions thoroughly. It should be made clear to them that a sample space diagram can either record two separate results in each cell or the outcome of the both results, such as adding them or finding the difference between them, or whatever is required by the particular question. They should make the decision about what to record themselves.

Differentiation

Work on combined events can be extended to more complex situations. For example, you could have two spinners, one numbered 8, 9, 10, 12, 15 and the other numbered 4, 6, 14, 18, 24, and outcomes such as highest common factor or lowest common multiple.

Plenary

Compare the results from the lesson starter with the theoretical probabilities that have been calculated in question 1 of Exercise 3F in the Pupil Book. Discuss any differences.

Incorporating exercise:	3G
Homework:	3.7
Example:	3.7

Key words
combined
   events
space diagram
tree diagram

Learning objective(s)

● use sample space diagrams and tree diagrams to work out the probability of combined events

Prior knowledge

Pupils should have completed Section 3.6 and be able to calculate the probability of two events happening at the same time through the use of sample space diagrams.

They should be confident in adding and multiplying simple fractions and, for probabilities that are given in decimal form, adding and multiplying decimals.

Starter

Give pupils some typical calculations involving sums and products of fractions. The pace and focus of the lesson can be needlessly derailed if confidence is lacking in this area.

Main teaching points

The main focus of this section is on tree diagrams. These are a way of allowing probabilities of two or more successive events to be calculated for events that are not necessarily equally likely. However, their potential untidiness makes them impractical for situations where each event has more than three outcomes.

Each event is depicted by a branch of the tree. The probability of any event happening is written on the corresponding branch.

The probability of any outcome is calculated by multiplying together the probabilities along the respective branches. Several of these probabilities may then be added if the question is requiring the probability of 'one outcome or another'. Pupils may find it useful to remember the rule: 'multiply *along* the branches, and add *down* the tree'.

If probabilities are given in fractional form, final answers should always be simplified wherever possible.

Common mistakes

Some pupils may decide to add probabilities along the branches, rather than multiply them. It is also important to make sure that pupils draw the initial tree diagram to a reasonable size and label each branch with its corresponding probability, irrespective of what the question is asking.

Differentiation

Lower achieving pupils can find difficulties in setting up a tree diagram if the outline is not drawn for them. They may need to look at more examples and questions to help them in this area.

Plenary

Compare tree diagrams with sample space diagrams from Section 3.6. Ask pupils what advantages or disadvantages one might have over the other. Can they come up with scenarios where it is clearly preferable to use tree diagrams instead of sample space diagrams, and vice versa?

Incorporating exercises:	3H, 3I, 3J	**Key words**
Homework:	3.8	and
Example:	3.8	independent events
		or

Learning objective(s)

● use the connectors 'and' and 'or' to find the probability of combined events

Prior knowledge

Pupils must know how to find the probability of combined events using sample space diagrams and tree diagrams. They should be familiar with mutually exclusive events, and know that if A and B are two such events, then $P(A \text{ or } B) = P(A) + P(B)$.

Starter

Can pupils solve a tree diagram question without the use of a tree diagram? For example, suppose a dice is thrown twice. What is the probability of scoring exactly one six from both throws?

How could we then work out the probability of scoring exactly one six from three throws?

Main teaching points

If two events A and B are such that the outcome of one does not affect the outcome of the other, then A and B are said to be independent events. Suppose, for example, a bag contains five red balls and four yellow balls and two balls are taken out, one after the other, without replacement. Then the probability of the second ball being red will depend on the colour of the first ball taken from the bag. Therefore, these events would not be independent. However, if the first ball was replaced before the second was taken, then the two events would be independent.

For any two independent events, A and B, we have that $P(A \text{ and } B) = P(A) \times P(B)$.

Some questions in Exercises 3H, 3I and 3J can be worked out using tree diagrams. However, these are impractical for cases involving a sequence of events that generate a large number of outcomes, such as rolling a die or tossing a coin more than two or three times. The calculation used is effectively the same, but relies on the words 'and' and 'or'. If pupils are asked to calculate the probability of a compound event, they should firstly think of the possibilities of how that event could happen, using the words 'and' and 'or'. Then, remembering '*and* means multiply and *or* means add', they should be able to write down the calculation necessary to solve the problem.

Many questions involving combined events ask for cases where at least one of the events occurs. In these cases, it is best to use $P(\text{at least one happens}) = 1 - P(\text{none happens})$. Questions of this type have already been met by pupils (see, for example, Homework 3.7 and Worked examples 3.7) without the method so far being formally taught.

Common mistakes

Pupils are often too quick to think of one possibility of a compound event happening, without considering all eventualities. For example, suppose a coin is tossed twice. What is the probability of getting a head and a tail? They should be writing P(head and tail) = P(head first and tail second *or* tail first and head second). Less able pupils are likely to plunge straight into P(head and tail) = $\frac{1}{2} \times \frac{1}{2}$, thus missing one or more valid outcomes. Encourage pupils to think of all the possibilities before the probabilities.

Differentiation

The vast majority of questions are grade A standard and so may prove inaccessible to a significant number of pupils. They may prefer to do some initial working with the help of a tree diagram, although this would not be practical for all questions in the Pupil Book.

Plenary

Ask pupils to define mutually exclusive events and independent events.

Pose the following questions:
- If two events A and B are mutually exclusive, what can be said about $P(A$ or $B)$?
- If two events A and B are independent, what can be said about $P(A$ and $B)$?
- Can pupils suggest two events which are/are not mutually exclusive?
- Can pupils suggest two events which are/are not independent?

Incorporating exercise:	3K
Homework:	3.9
Example:	3.9

Key words
conditional
probability

Learning objective(s)

● work out the probability of combined events when the probabilities change after each event

Prior knowledge

Pupils should have covered Sections 3.7 and 3.8 and be confident in calculating probabilities of compound events with and without tree diagrams.

Starter

Go through question 1 from Worked examples 3.9 with pupils, but initially using a tree diagram. This way, the 'dependent' probabilities attached to the second column branches may be seen more clearly and discussed.

Main teaching points

This section on conditional probability relates to dependent events in comparison to the previous section on independent events. The questions involved are almost identical in style, and situations are generally similar, except for changes that have the effect of altering probabilities relating to sequences of events. Pupils will already have come across some relatively simple conditional probability problems in the sections on tree diagrams and independent events, this section builds on that work by introducing pupils to more challenging problems.

Common mistakes

As in the previous section, pupils can miss out on all the possibilities that are required to answer a particular question. Since this topic mainly concerns itself with conditional probability and dependent events, there is also always the chance that individual events will be treated as independent events rather than the dependent events that are now more likely to be seen. In any case, more able pupils should quickly be able to develop a better awareness of dependent events and their associated probabilities.

Differentiation

Conditional probability is one of only a few topics aimed at discriminating between the most able. A clear understanding of the main aspects of probability covered so far is desirable. Less able pupils (relatively speaking) should be able to understand the easier type of question in this section – the suggested starter should be accessible, for instance. Only the most able will be able to understand and produce correct solutions for problems involving dependent events without the aid of tree diagrams.

Plenary

Try pupils on a mix of questions to cover various aspects of probability covered in the chapter.

- Two cards are drawn one at a time from a pack of cards. The cards are replaced each time. What is the probability that at least one of them is a spade? (Answer: $\frac{7}{16}$)

- A fair dice is thrown twice. What is the probability of getting at least one six? (Answer: $\frac{11}{36}$ (0.306))

- The dice is thrown three times. What is the probability of getting at least one six now? (Answer: 0.421)
- The dice is thrown four times. What is the probability of getting at least one six? (Answer: 0.518)

- The dice is thrown n times. What is the probability of getting at least one six? (Answer: $1 - (\frac{5}{6})^n$)

- There are five socks in a drawer, of which three are blue and two are black? You take out two socks. What is the probability that:
  a both socks are blue?
  b both socks are black?
  c you get a pair of socks?
  d at least one of the socks is blue?

  (Answers: **a** $\frac{3}{10}$ **b** $\frac{1}{10}$ **c** $\frac{2}{5}$ **d** $\frac{9}{10}$)

Number

Overview

4.1 Solving real problems
4.2 Division by decimals
4.3 Estimation
Rounding off to significant figures
Multiplying and dividing by multiples of 10,
approximation of calculations and sensible
rounding
4.4 Multiples, factors and prime numbers
4.5 Prime factors, LCM and HCF
4.6 Negative numbers

This chapter covers a variety of issues – starting with solving real-life problems without a calculator. It then moves on to dividing by decimals, and then to estimating various calculations. Section 4.5 builds on the revised skills of Section 4.4, so that pupils can write a number in prime factor form and find least common multiples (LCMs) and highest common factors (HCFs). The last section deals with negative numbers and the four rules $(+, -, \times, \div)$, with emphasis on multiplying and dividing.

Context

Solving real-life problems, unsurprisingly, is aimed at real life. Estimation is very useful in checking that any calculation is "about right".

AQA B references

AO2 Number and algebra: Numbers and the number system

4.4, 4.5   2.2a "… use the concepts and vocabulary of factor (divisor), multiple, common factor, highest common factor, least common multiple, prime number and prime factor decomposition"

AO2 Number and algebra: Calculations

4.2, 4.3, 4.5, 4.6   2.3a "multiply or divide any number by powers of 10, and any positive number by a number between 0 and 1; find the prime factor decomposition of positive integers; … multiply and divide by a negative number; …"
4.3   2.3h "round to a given number of significant figures; …"

AO2 Number and algebra: Solving numerical problems

4.1, 4.3   2.4b "check and estimate answers to problems; select and justify appropriate degrees of accuracy for answers to problems; …"

Route mapping

Exercise	D	C	B	A	A*
A	all				
B	all				
C	1–3	4–5			
D	all				
E	1–9	10–13			
F	1–3	4–8			
G	1–8	9	10–11		
H		all			
I		all			
J	all				
K	all				

Answers to diagnostic Check-in test

1 a 11 256   **b** 27   **c** 49

2 a Any three from: 7, 14, 21, 28, 35, 42, …   **b** Any three from: 2, 3, 5, 7, 11, 13, 17, 19, …

   c Any three from: 1, 4, 9, 16, 25, 36, 49, 64, 81   **d** Any three from: 1, 2, 3, 5, 6, 10, 15, 30

3 a 30   **b** 50   **c** 18

4 a £130.44   **b** £8.75

5 a 217.5   **b** 6350   **c** 8110   **d** 56

Answer all of the following questions WITHOUT using a calculator.

1 Work out the following.

   a 42×268 **b** $837 \div 31$ **c** $(2 + 5)^2$

2 Write down the following:

   a 3 multiples of 7

   b 3 prime numbers

   c 3 square numbers less than 90

   d 3 factors of 30

3 Use your skill with BODMAS to work out the following.

   a $5 + 5 \times 5 =$

   b $(5 + 5) \times 5 =$

   c $5 + 4^2 - 3 =$

4 Work out the following.

   a $£21.74 \times 6$ **b** $£52.50 \div 6$

5 Write down the answers to the following.

   a $21.75 \times 10 =$ **b** $6.35 \times 1000 =$

   c $81.1 \times 100 =$ **d** $0.056 \times 1000 =$

Incorporating exercise: 4A
Homework: 4.1
Examples: 4.1

Key words
long division   strategy
long
    multiplication

Learning objective(s)

● using arithmetic to solve more complex problems

Prior knowledge

Pupils must be able to multiply numbers such as 24×36, 162×78, $£24.50 \times 15$, and divide numbers such as $1035 \div 55$, without a calculator.

Starter

Practise multiplication and division without a calculator. Make sure some of the questions involve money (to 2 dp) or measurement (to 1 or 2 dp).

For example, 22×38 (836), 204×72 (14 688), $£32.15 \times 12$ (£385.80), $356\,km \div 5$ (71.2 km), $886\,m \div 14$ (63.29 m to 2 dp).

Main teaching points

Understanding the question is essential. Get the pupils to read the question several times before starting.

Common mistakes

Rushing, and so not understanding properly. This leads to mistakes such as multiplying instead of dividing, etc. This is often not picked up by pupils, as they have not estimated first to get a rough idea of what their answer should be.

Differentiation

This is the first chapter in the Higher tier GCSE book – if pupils really struggle with this and the next few sections, thought must be given to the ability of the pupils: would they be better off doing the Foundation GCSE?

Plenary

Give out Homework 4.1. Ask the pupils to read the questions carefully.
Read out question 1 and ask how it is to be solved.
Discuss the remaining questions, and the skills needed to complete them successfully.

Incorporating exercise:	4B	**Key words**	
Homework:	4.2	decimal places	integer
Example:	4.2	decimal point	

Learning objective(s)

● divide by decimals by changing the problem so you divide by an integer

Prior knowledge

Pupils must be able to multiply decimal numbers by 10, 100, 1000, etc., and be able to use pencil and paper methods for division.

Starter

Give several divisions questions with integer values and integer answers.
For example, $342 \div 3$ (114), $8561 \div 7$ (1223), $2784 \div 12$ (232).

Main teaching points

Make sure that pupils understand that $3.6 \div 0.2$ and $36 \div 2$ are essentially the same (that is they have the same values, but different appearance).

Common mistakes

Multiplying one part of the division by (for example) 100, but not the other part, or by multiplying the two parts by different numbers (for example, 10 and 1000).
Just making each part a whole number.

Plenary

Ask the class to divide 36 by $\underbrace{0.1, 0.2, 0.3, 0.4, 0.5, 0.6, 0.8, 0.9,}_{>36} \underbrace{1.0}_{=36} \underbrace{1.2, 1.8, 3.6.}_{<36}$

If there is time, repeat with 3.6.

		Key words	
Incorporating exercise:	4C	approximate	significant
Homework:	4.3a	estimation	figures
Examples:	4.3a		

Learning objective(s)

● use estimation to find approximate answers to numerical calculations

Prior knowledge

Being able to round off to a given number of decimal places (which is an easier skill to master), will help the pupils learn this skill.

Starter

Pose the following questions:
● Round 23.468231 to **a** 1 dp **b** 2 dp **c** 3 dp **d** 4 dp
  (**a** 23.5 **b** 23.47 **c** 23.468 **d** 23.4682)
● Round 145.9952 to **a** 1 dp **b** 2 dp **c** 3 dp
  (**a** 146.0 **b** 146.00 **c** 145.995)
● I am 35 years old. Discuss how old I actually could be.

Main teaching points

When **estimating** (a key word in exams) an answer to a calculation, we round off all numbers to 1 significant figure, that is, one number, and then as many zeros as we need to keep the estimation the same order of magnitude as the original number.

Emphasise that 0, 1, 2, 3 or 4 as the 'next digit' means leave the digit on the left alone, and when 5, 6, 7, 8 or 9 is the 'next digit', the digit on the left goes up by one.

Common mistakes

These are best shown with two examples, both rounded to 2 significant figures:
  275 986 = 28, rather than 280 000, (the zeros are forgotten)
  0.00037104 = 0.00 or 37, rather than 0.00037 (2 dp instead of 2 sf, or zeros forgotten)

Differentiation

Any pupil who finds this difficult must be reminded to make sure the rounded off number is just that – rounded off, not drastically changed.

Plenary

Write on the board:

571.8 to 1 sf = 6   } Play Devil's advocate in arguing that
0.003 276 to 2 sf = 0.00   } these are OK – get the pupils to
25.37 to 1 sf = 30.00   } disprove your answers.

Incorporating exercises:	4D, 4E, 4F
Homework:	4.3b
Examples:	4.3b

Key words

approximate	significant
estimation	figures

Learning objective(s)

● use estimation to find approximate answers to numerical calculations

Prior knowledge

Good mental arithmetic skills of multiplication and division are useful, as are problem-solving skills in real-life type maths questions.

Starter

Ask the pupils a selection of the following mental arithmetic questions:
● Multiplication questions
  For example, any questions using the times tables.
● Division questions, some with 1 decimal place answers.
  For example, $72 \div 8$ (9), $42 \div 6$ (7), $10 \div 4$ (2.5), $78 \div 10$ (7.8), $16 \div 20$ (0.8).
● BODMAS questions to refresh their memories.
  For example, $4 + 3 \times 5$ (19), $12 - 3^2$ (3), $(2 + 7) \div 3$ (3).

Main teaching points

Emphasise that with both multiplying and dividing numbers by different powers of 10 and approximation of calculations, the idea is to make life as easy as possible. If they feel tempted to use a calculator it may be that they have not made the question easy enough to do without a calculator. When rounding, each number must be rounded in the context of the question, but often just rounding to the same significant figure (or most significant figure) as the original numbers in the question is enough.

Common mistakes

Pupils often get into a muddle with the power of 10 in the answer.

When approximating calculations, pupils often try to be too accurate, rounding to more than 1 significant figure, or not rounding at all and using a calculator, assuming a totally correct answer is much better than an approximate answer. Similarly, when rounding off answers, pupils try to be too accurate, not appreciating the accuracy needed in real life situations – some common sense is needed here.

Plenaries

● Go over methods for multiplying numbers by different powers of 10.
  For example, to multiply 200 by 0.2, 'take' a 0 from the 200, and 'give' it to the 0.2 ($\div 200$ by 10 and $\times 0.2$ by 10), leaving $20 \times 2 = 40$.
● Go over methods for dividing numbers by different powers of 10.
  For example, to divide 200 by 0.2, multiply both numbers by 10 to get $2000 \div 2$ (= 1000).
● Ask, "What is an approximate answer to $\dfrac{35961}{91.4}$?" Give the options $\dfrac{40000}{90}$ or $\dfrac{36000}{90}$.
  Discuss which would be easiest to do, and why.

Incorporating exercise:	4G
Homework:	4.4
Examples:	4.4

Key words

factor	square number
multiple	triangular
prime number	number

Learning objective(s)

● revise multiples, factors, prime numbers, square numbers, triangular numbers, square roots, cubes and cube roots
● estimate using square roots and cube roots

Prior knowledge

Most of the learning objectives have been covered before, some several times, in previous years. This is just a quick reminder in order to use these skills later.

Starter

Ask the pupils to explain what multiples, factors, prime numbers, square numbers, triangular numbers, square roots, cubes and cube roots are. If they can't explain, can they give an example or two?

Main teaching points

Emphasise that these ideas have been covered before; they are being done again because they will be tested at GCSE level, and they are used to help with other mathematical skills.

Common mistakes

Missing out 1 as a factor, and including 1 as a prime number.

Plenary

Write the words multiple, factor, prime number, square number, triangular number, square root, cube and cube root on the board. Ask each pupil in turn to give an example (or examples) of the word you are pointing at – occasionally ask them to justify their choice or say that you're not keen on that number and ask for another.

Incorporating exercises:	4H, 4I
Homework:	4.5
Examples:	4.5

Key words
highest common   prime factor
   factor (HCF)
least common
   multiple (LCM)

Learning objective(s)

● write a number as a product of its prime factors
● find the least common multiple (LCM) and highest common factor (HCF) of two numbers

Prior knowledge

Pupils must be happy dealing with prime numbers, multiples and factors.

Starter

Ask the pupils for the first 10 prime numbers.
Ask the pupils for any 5 multiples of 4, of 8 and of 12.
Ask the pupils for all the factors of 20, 50 and 60.

Main teaching points

Most pupils find the factor 'ladder' the safest way to find the prime factors of a number.

To find the LCM of small numbers it is often quickest to write out the multiples, and pick the smallest one. With large numbers encourage the use of prime factors using the 'ladder' again – see Worked examples 1.5, question 2.

When finding the HCF of small numbers, always write out all the factors and pick the largest one. With large numbers, again use prime factors and the 'ladder' method – see Worked examples 1.5, question 3.

Common mistakes

When finding prime factors, a common mistake is to not use prime numbers.
For example, $36 = 2 \times 2 \times 9$.

Plenary

Ask a volunteer to give each step in the method for finding the LCM of 96 and 72 by using a factor ladder (288). What is the HCF of 96 and 72? (24)

Do the same for 84 and 420 (LCM = 420, HCF = 84).

     Module 3: Number

Incorporating exercises:	4J, 4K	**Key words**	
Homework:	4.6	negative	positive
Examples:	4.6		

Learning objective(s)

- multiply and divide with positive and negative numbers

Prior knowledge

Pupils must be confident with the order of operations (BODMAS).

Starter

Put some multiplications on the board for pupils to work out with a calculator.
For example, 3×6, 3×-6, -3×6, -3×-6

Ask what rules can be found regarding positive and negative numbers when multiplying.
- When the signs of the numbers are the same, the answer is positive.
- When the signs of the numbers are different, the answer is negative.

Do these rules also apply for division?

Main teaching points

Keep asking the pupils what the rules are. These must be learnt. Also, check that the pupils know how to input negative numbers into their calculators.

Common mistakes

Applying the rule for multiplying negative numbers to addition and subtraction problems. For example, $-3 - 4 = +7$.

Pupils often get confused with problems such as 3×-4. What does a $\times$ and a $-$ make?

Plenary

Ask each pupil in turn simple mental questions (no writing allowed).
For example, 4×-3, $8 \div -2$, $4 - 9$, etc.

Fractions and percentages

Section 5.1 shows how to find one quantity as a fraction of another. Sections 5.2, 5.3, and 5.4 deal with the four rules $(+, -, \times, \div)$ and fractions. Sections 5.5 to 5.8 focus on all the common GCSE techniques at this higher level, that is, percentage increase and decrease, expressing one quantity as a percentage of another, compound interest, repeated percentage change, and reverse percentages.

Context

Fractions are used less and less, the more the metric system is used, but there will always be instances where using fractions is easier than their decimal equivalents. Percentages impinge on much of our everyday life – VAT, profit and loss, interest rates, income tax . . . .

AQA B references

AO2 Number and algebra: Calculations

5.1, 5.2   2.3c "calculate a given fraction of a given quantity, expressing the answer as a fraction; express a given number as a fraction of another; add and subtract fractions by writing them with a common denominator; . . ."

5.3, 5.4   2.3d ". . . multiply and divide a given fraction . . . by a unit fraction and by a general fraction"

5.5, 5.7, 5.8   2.3e ". . . understand the multiplicative nature of percentages as operators; calculate an original amount when given the transformed amount after a percentage change; reverse percentage problems"

5.5–5.8   2.3l "solve percentage problems, including percentage increase and decrease, and reverse percentages"

5.7   2.3n "represent repeated proportional change using a multiplier raised to a power"

5.8   2.3w "use calculators for reverse percentage calculations by doing an appropriate division"

AO2 Number and algebra: Solving numerical problems

5.5–5.8   2.4a "draw on their knowledge of operations and inverse operations . . . in order to select and use suitable strategies and techniques to solve problems and word problems, including those involving . . . repeated proportional change, fractions, percentages and reverse percentages . . ."

Route mapping

Exercise	D	C	B	A	A*
A	1–4	5–8			
B	1–8	9			
C	1–3	4–7	8		
D	1–4	5			
E	1–7	8	9		
F	1–9		10	11	
G	1–4	5–9			
H		1–7	8–9	10–11	
I		1	2–9	10–11	

Answers to diagnostic Check-in test

1 a $\frac{1}{4}$ **b** $\frac{2}{7}$ **c** $\frac{1}{3}$

2 21

3 a 50%, 0.5   **b** 30%, 0.3   **c** 96%, 0.96   **d** $\frac{1}{5}$, 0.2   **e** $\frac{3}{4}$, 0.75

   f $\frac{2}{25}$, 0.08   **g** $\frac{9}{10}$, 90%   **h** $\frac{4}{25}$, 16%   **i** $\frac{7}{40}$, 17.5%

1 Cancel down these fractions to their simplest form.

a $\dfrac{5}{20} =$ **b** $\dfrac{4}{14} =$ **c** $\dfrac{40}{120} =$

2 Find the lowest common multiple (LCM) of 3 and 7.

3 Complete this table of equivalent fractions, percentages and decimals.

	Fraction	Percentage	Decimal
a	$\frac{1}{2}$		
b	$\frac{3}{10}$		
c	$\frac{24}{25}$		
d		20%	
e		75%	
f		8%	
g			0.9
h			0.16
i			0.175

Incorporating exercise:	5A	Key words
Homework:	5.1	cancel
Example:	5.1	fraction

Learning objective(s)

● find one quantity as a fraction of another

Prior knowledge

Pupils should be able to do simple division.

Starter

Use a metre rule or draw one on the board. Show 20 cm on the rule and ask what fraction this is of the whole rule. Can this be simplified? Repeat with other numbers of centimetres. Now ask what fraction 20 cm is of 2 m, etc. Look at other measures, such as money or time.

Main teaching points

The first number in the problem becomes the numerator of the fraction and the second the denominator. Answers should be simplified if possible.

Common mistakes

Incorrect cancelling down.

Plenary

Use a circle and ask questions about angles of sectors, for example what fraction of a whole circle is 36°?

Incorporating exercise:	5B
Homework:	5.2
Examples:	5.2

Key words

denominator	lowest common
equivalent	denominator
fraction	

Learning objective(s)

• add and subtract fractions with different denominators

Prior knowledge

Pupils need the skills of being able to find the LCM of 2 or 3 numbers, and being able to convert mixed numbers into improper fractions.

Starter

Ask pupils to find the LCM of **a** 2 and 3 **b** 2 and 4 **c** 2 and 5 **d** 5 and 7.
Ask pupils to change $3\frac{1}{5}$ into an improper fraction.

Main teaching points

Emphasise that the fractions are changed into *equivalent fractions* when they are being changed to make the denominators the same, in order that they can be added or subtracted.

Common mistakes

Not converting to equivalent fractions to make the denominators the same – simply adding or subtracting the numerators and denominators. Changing the denominators, but forgetting to change the numerators.

Plenary

Write this calculation on the board: $\frac{1}{2}+\frac{1}{3}-\frac{1}{4}-\frac{1}{6}$

Discuss strategies of what to look for and how to make sure the answer is found most efficiently.

Incorporating exercise:	5C	Key word
Homework:	5.3	cancel
Examples:	5.3	denominator
		numerator

Learning objective(s)

● multiply fractions

Prior knowledge

Pupils should be able to cancel fractions and know the multiplication tables up to 10×10.

Starter

Pose the following questions:

a Cancel down $\frac{20}{35}$.

b Cancel down $\frac{4 \times 5}{7 \times 5}$. Is this the same as **a**?

Yes – notice that the 5s cancel each other out.

Repeat with:

c $\frac{16}{20}$ and $\frac{4 \times 4}{5 \times 4}$

d $\frac{12}{30}$ and $\frac{6 \times 2}{6 \times 5}$

Main teaching points

Cancelling (as in the starter) is not essential, but it is a very useful skill for making this type of question much easier, as the numbers involved are therefore smaller.

Common mistakes

Incorrect multiplications, for example $3 \times 3 = 6$, and converting a mixed number to an improper fraction incorrectly.

Plenary

Write this calculation on the board: $1\frac{2}{5} \times 2\frac{3}{7}$

Ask pupils for the two main methods of multiplying these two fractions, that is, once the mixed numbers are converted to improper fractions, would they cancel first or last? Which do they prefer and why?

Incorporating exercise:	5D	Key word
Homework:	5.4	reciprocal
Examples:	5.4	

Learning objective(s)

- divide by fractions

Prior knowledge

Pupils must be able to multiply fractions.

Starter

Give the pupils a variety of questions involving multiplication of fractions.
For example, $\frac{1}{2} \times \frac{1}{5}$ and $1\frac{3}{5} \times \frac{5}{8}$.

Main teaching points

For some pupils showing that to divide by a fraction you simply multiply by the reciprocal, is not enough. You may have to explain *why* it works; giving simple examples such as $4 \div \frac{1}{2} = 8$ will help the pupils understand why.

Common mistakes

A range of mistakes can occur, but the two main ones are **a** simply dividing the numerators and denominators (as in multiplying) and **b** finding the reciprocal of the wrong fraction or of both fractions.

Plenary

With questions such as $\frac{4}{5} \div \frac{1}{2}$ and $1\frac{1}{4} \div 3\frac{1}{3}$, get pupils to give you step-by-step instructions on how to find the answers.

Incorporating exercises:	5E, 5F
Homework:	5.5
Examples:	5.5

Key words

multiplier	percentage increase
percentage decrease	

Learning objective(s)

- calculate percentage increases and decreases

Prior knowledge

Pupils should know, for example, that $5\% = \frac{5}{100}$.

Starter

Ask the class, "If Alice gets £10 pocket money per week now, how much will she get after a 50% rise in her pocket money? How can we show the workings?"
Pupils will probably say: 50% = £5, so £10 + £5 = £15.
Explain the method of using the multiplier 1.5 for 150%, so £10 × 1.5 = £15.
Repeat for a 50% reduction in pocket money.

Main teaching points

For all but the weakest higher tier pupils, insist on the multiplier method whenever possible – it is much quicker and can save a lot of time in an exam.

Common mistakes

For an increase of say 5%, using the multiplier as 1.5 rather than 1.05. Also, correctly using the multiplier, but then adding it on to the original value.

Differentiation

Lower achieving pupils may not be able to understand the multiplier method, so let them take the easier but slower method of finding the percentage increase or decrease then adding or subtracting it from the original amount.

Plenary

Using a calculator, and without writing anything down ask pupils to:
- increase £75 by 75%   (£131.25)
- decrease £75 by 75%   (£18.75)

Incorporating exercise:	5G	Key words
Homework:	5.6	percentage gain percentage loss
Examples:	5.6	

Learning objective(s)

● express one quantity as a percentage of another

Prior knowledge

Being able to write one quantity as a fraction of another quantity is an immediate advantage to understanding this section.

Starter

Ask the class, "What is £10 as a fraction of £40?" ($\frac{1}{4}$)
"What is £10 as a percentage of £40?" (25%)
Discuss how they calculated their answers.

Main teaching points

When expressing one quantity as a percentage of another, the original amount is the denominator of the fraction.

For profit and loss questions, pupils should *learn* the formula: $\text{profit}/\text{loss} = \dfrac{\text{difference}}{\text{original}} \times 100\%$

Common mistakes

Not using the original value as the denominator, but the new value after the increase or decrease.

Plenary

Pose the following questions:
● Express 40p as a percentage of £4 – ask what the obvious mistakes would be.
● Sam buys a painting for £1500 and sells it for £2000. What is Sam's percentage profit?
● If Sam had sold the painting for £1200, what would the percentage loss be?

    Module 3: Number

Incorporating exercise:	5H	Key words	
Homework:	5.7	annual rate	principal
Examples:	5.7	multiplier	

Learning objective(s)

- calculate compound interest
- solve problems involving repeated percentage change

Prior knowledge

Pupils should be able to increase an amount by a percentage, preferably using the multiplier method.

Starter

Give the class the following problem:
Craig invests £100 in a savings account, which earns him 5% interest per annum.
How much does Craig have in the account at the end of the first year? (£105)
How much does Craig have in the account at the end of the second year? (£110.25)
How much does Craig have in the account at the end of the third year? (£115.76)

Discuss a quicker method to find out how much Craig had at the end of the third year?
(100×1.05^3)

Main teaching points

In order to speed up calculations for compound interest, it is important that pupils can understand and use the multiplier method from the previous section.

Get the pupils to learn and use the compound interest formula:

$$A = P\left(1 + \frac{r}{100}\right)^n$$

or, in multiplier form: $A = P(1 + x)^n$ where $(1 + x)$ is the multiplier.

Get the pupils to understand that these two formulae are identical, they appear different because of the form in which they are written.

Common mistakes

When doing the year-by-year method, premature rounding or incorrect rounding often leads to incorrect answers. When using the compound interest formula for *depreciation*, the pupils often forget that the ' + r%' becomes a '− r%'. Another common mistake is not reading the question properly – does the question ask for the total amount in the bank after n years, or for the total interest received after n years?

Differentiation

Lower ability pupils may only be able to work these questions out by calculating 1 year at a time. Encourage the higher ability pupils to use the compound interest formula.

Plenary

Pose these problems to the class:
- Mr Casio invests £1 000 000 for 5 years at an annual compound interest rate of 5%. By how much does his money grow?
- Miss Casio buys a car for £100 000 which depreciates by 5% per year. By how much has the car depreciated after 5 years?
- Mrs Casio invests £1 000 000 at a rate of 5% per annum, compound interest. How many years will it take for this money to reach at least £1.5 million?

Incorporating exercise:	5l	**Key words**	
Homework:	5.8	multiplier	unitary method
Examples:	5.8	original amount	

Learning objective(s)

●   calculate the original amount after you know the result of a percentage increase or decrease

Prior knowledge

Pupils should be able to work out a multiplier, that is add to 100 if an increase, or subtract from 100 if a decrease, and express the result as a decimal. They should be able to work out the percentage of an amount.

Starter

Ask, "What is 30% of £200?"   (£60)

$200 \times 0.3 = 60$ ◄———   Write on the board and leave there.

Ask, "If you were told that 30% of an amount was worth £60, how would you find the original amount?"

Reverse the process: ———► $60 \div 0.3 = 200$

Main teaching points

The quickest method is undoubtedly using the multiplier method, but many pupils will prefer a unitary method approach (see Worked examples 5.8).

Common mistakes

Not using the correct multiplier.

Differentiation

Try the multiplier approach first and fall back on the unitary method approach if needed. Alternatively, just use the unitary method approach.

Plenary

Get the pupils to help you with a few questions such as:

●   VAT: An MP3 player costs £176.25 including VAT, how much would it have cost without VAT of 17.5%?   (£150)

●   Salary: Pam now earns £21 840 after a 4% pay rise. How much did she earn before her pay rise?   (£21 000)

●   Weight: Paul keeps a diary of his weight. He lost 6% of his body weight in 1 month, then gained 6% of his new body weight during the following month. Did he gain or lose weight overall? What was the percentage change? If he weighs 75 kg now, what was his weight to begin with?   (Lose by 0.36%; 75.3 kg)

Ratios and proportion

Overview

6.1 Ratio
6.2 Speed, time and distance
6.3 Direct proportion problems
6.4 Best buys
6.5 Density

In this chapter, pupils will learn what a ratio is, how to simplify a ratio and express it as a fraction. They will divide amounts into given ratios and solve problems. Pupils will then look at speed-distance-time problems, and then study direct proportion using the unitary method, which leads onto best buys. The chapter concludes with a section on density, mass, and volume problems.

Context

There are many practical uses of ratio and proportion; the chapter covers the most commonly used. This includes recipes, mixing paints, looking at speed over time, density, and especially looking at best buys.

AQA B references

AO2 Number and algebra: Using and applying number and algebra

6.1–6.5   2.1a "select and use appropriate and efficient techniques and strategies to solve problems of increasing complexity, involving numerical and algebraic manipulation"

AO2 Number and algebra: Numbers and the number system

6.1   2.2f "use ratio notation, including reduction to its simplest form and its various links to fraction notation"

AO2 Number and algebra: Calculations

6.1   2.3f "divide a quantity in a given ratio"

AO2 Number and algebra: Solving numerical problems

6.1–6.5   2.4a "draw on their knowledge of operations and inverse operations ... in order to select and use suitable strategies and techniques to solve problems and word problems, including those involving ratio and proportion ... and compound measures ..."

AO2 Number and algebra: Equations, formulae and identities

6.1–6.5   2.5f "set up simple equations; solve simple equations ..."

Route mapping

Exercise	D	C	B	A	A*
A	all				
B		all			
C		all			
D	1–8	9–12			
E	all				
F	all				
G			all		

Answers to diagnostic Check-in test

1 a $\dfrac{1}{2}$ **b** $\dfrac{1}{4}$ **c** $\dfrac{2}{3}$ **d** $\dfrac{5}{7}$

2 £36   **b** 2.5 cm   **c** 224 litres   **d** 72 kg

3 2.5 hours

      Module 3: Number

1 Simplify these fractions by cancelling.

a $\dfrac{15}{30} =$

b $\dfrac{18}{72} =$

c $\dfrac{44}{66} =$

d $\dfrac{35}{49} =$

2 Find the following quantities.

a $\dfrac{3}{4}$ of £48

b $\dfrac{1}{8}$ of 20 cm

c $\dfrac{2}{3}$ of 336 litres

d $\dfrac{4}{5}$ of 90 kg

3 If I am in a car travelling at 60 mph, how long will it take me to drive 150 miles?

Incorporating exercises:	6A, 6B, 6C	**Key words**	
Homework:	6.1a, 6.1b, 6.1c	common units	ratio
Examples:	6.1a, 6.1b, 6.1c	fraction	

Learning objective(s)

- simplify a ratio
- express a ratio as a fraction
- divide amounts into given ratios
- complete calculations from a given ratio and partial information

Prior knowledge

Pupils should know the times tables up to 10×10, how to cancel fractions, how to find a fraction of a quantity, and how to multiply and divide both with and without a calculator.

Starters

- Give the pupils fractions to cancel, for example $\frac{5}{15}$, $\frac{7}{14}$, $\frac{10}{15}$, $\frac{4}{10}$, $\frac{15}{25}$.
- Practise converting between metric units by asking quick-fire questions around the class.
- Give the pupils divisions, such as $500\,g \div 10$, $5\,kg \div 10$, $5\,hrs \div 10$, $5\,hrs \div 3$, $£5 \div 25$.
- Put pairs of fractions on the board, with either the numerator or denominator from one fraction missing. Ask for the missing number to make the fractions equivalent.

  For example, $\frac{2}{3} = \frac{80}{?}$

Main teaching points

Pupils need to know that a ratio is a way of comparing the sizes of two or more quantities. The quantities must be of the same units, for example cm to cm. If the units are different then one must be converted before simplifying. This is because a ratio does not have units. A ratio can be expressed as a fraction. The denominator of the fraction is obtained by adding the number of parts in the ratio.

To divide (split up) amounts into given ratios, find the total number of shares (or parts). Divide by this and then multiply by the number of shares each receives. For example, when dividing 500 in the ratio of $2 : 6$, divide 500 by 8 and then multiply the answer by 2 or 6 to find each share.

Common mistakes

Not changing to a common unit. In expressing $3 : 5$ as a fraction, simply writing this as $\frac{3}{5}$.

Differentiation

In the section 'Dividing amounts into given ratios' in the Pupil Book, give lower achieving pupils just the Example 3 version.

Plenary

For each section, check understanding by going through the last two questions in Exercises 6A, 6B and 6C.

Incorporating exercise:	6D	**Key words**	
Homework:	6.2	distance	time
Examples:	6.2	speed	

Learning objective(s)

- recognise the relationship between speed, distance and time
- calculate average speed from distance and time
- calculate distance travelled from the speed and the time
- calculate the time taken on a journey from the speed and the distance

Prior knowledge

Pupils must know how to multiply and divide.

Starter

Talk about the different measurements used when dealing with speed, distance and time. Go over the meaning of mph, kph, km/h, m/s.

Main teaching points

The simplest way to explain the relationship between speed, distance and time is to think of miles *per* hour.

Pupils will be familiar with this, even if they have not met it formally before. Once this is written as $\frac{miles}{hour}$, it is usually easy for pupils to see that speed is how quickly one goes for a given distance over a given time.

Speed is therefore distance divided by time. We can write this as a rule (or formula):

$$speed = \frac{distance}{time}$$

Show the pupils how to rearrange the rule to find time and distance. Using simple examples that pupils can work out in their heads will help with forming the general rules.

$$speed = \frac{distance}{time} \qquad time = \frac{distance}{speed} \qquad distance = speed \times time$$

Explain that the speed that is being calculated is an average for the journey time.

Common mistakes

Confusing decimal parts of an hour with hours and minutes, for example writing 3.50 h as 3 h 50 min.

Plenary

Simple, quick-fire questions that pupils can work out in their heads will check understanding.
For example, what is my speed if I travel 12 miles in:

a $\frac{1}{2}$ hour, **b** 2 hours, **c** 3 hours, **d** 4 hours, **e** 6 hours, **f** 12 hours, **g** 24 hours?

(**a** 24 mph, **b** 6 mph, **c** 4 mph, **d** 3 mph, **e** 2 mph, **f** 1 mph, **g** $\frac{1}{2}$ mph)

Incorporating exercise:	6E	**Key words**	
Homework:	6.3	direct	unitary
Examples:	6.3	proportion	method

Learning objective(s)

- recognise a direct proportion problem
- solve a problem involving direct proportion

Starter

Say to your pupils, "I have five blue counters, three green counters and two red counters." Ask for some statements that use ratio and proportion.

Main teaching points

Two quantities are in direct proportion if their ratio remains constant when the quantities increase or decrease. The easiest way to solve problems of direct proportion is to use what is known as the unitary method. This is where the value of 'one' is established. This is the constant factor that links the two quantities.

Plenary

Write a simple recipe on the board for six people. Ask the pupils what would be needed for one person, two people, 12 people, etc.

For example, Macaroni and Broccoli Cheese (serves 6)

240 g macaroni
75 g butter
75 g flour
600 ml milk
240 g cheese
300 g broccoli
45 ml breadcrumbs

Incorporating exercise:	6F	Key words
Homework:	6.4	best buy
Examples:	6.4	

Learning objective(s)

- find the cost per unit weight
- find the weight per unit cost
- find which product is the cheaper

Prior knowledge

Pupils must know how to multiply and divide.

Starter

Ask quick questions to revise pupils' knowledge of proportion. For example:
- If one item costs 35p how much will four items cost? (£1.40)
- If six items cost 90p, how much for one, 12, three, etc.? (15p, £1.80, 45p)

Main teaching points

Any pupils who shop in supermarkets will have met best buys before. Most supermarkets now show a unit or 100 g cost alongside the total cost of a product. Make sure that your pupils really read the instruction, for example "find the cost per unit weight". The key to answering these questions is in the wording. Per means divide, so cost per unit weight means cost divided by weight. Similarly weight per unit cost means weight divided by cost. Once this is understood solving these kinds of problems becomes simpler.

The smaller the unit cost the better value an item is. The greater the weight per unit cost the better value it is.

Common mistakes

Not making the units the same for each item. Not realising which is the best buy once the calculations are complete.

Plenary

Ask the pupils if they think the larger quantities will be proportionally cheaper than the smaller. Why might it not always be the best option to buy the biggest?

6.5 Density

Incorporating exercise:	6G
Homework:	6.5
Examples:	6.5

Key words
density   volume
mass

Learning objective(s)

● solve problems involving density

Prior knowledge

For this exercise, weight and mass are considered the same thing.

Starter

Put the distance–speed–time triangle on the board, and ask a few basic questions such as:
● when speed = 30 mph, time = 2 hours, find the distance (60 miles)
● when distance = 20 m, time = 4 s, find the speed (5 m/s)
● when distance = 100 m, speed = 25 m/s, find the time (4 s)

Main teaching points

If the pupils are used to the speed–distance–time triangle, then the density–mass–volume triangle should be easily accessible.

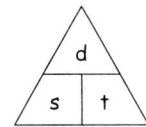

Common mistakes

Using the wrong formula, for example $d = m \times v$.

Plenary

Ask a pupil to draw the mass–density–volume triangle on the board.
Ask quick-fire, simple questions, using all versions of the formula.
For example, find the density, if mass = 2 kg and volume = 10 cm^3 (0.2 kg/cm^3).
Find the mass, if density = 2 g/cm^3 and volume = 25 cm^3 (50 g).
Find the volume, if mass = 50 g and density = 5 g/m^3 (10 m^3).

Overview

7.1 Indices
7.2 Standard form
7.3 Reciprocals and rational numbers
7.4 Surds

This chapter is broken down into four sections; the main parts being indices and standard form. Section 7.3 is concerned with reciprocals and rational numbers (including terminating and recurring decimals), while Section 7.4 deals with surds.

Context

Standard form is often used to express very large or very small numbers, particularly in aspects of science, for example space travel, microorganisms, etc.

AQA B references

AO2 Number and algebra: Calculations

7.1	2.3a "... use index laws to simplify and calculate the value of numerical expressions involving multiplication and division of integer, fractional and negative powers; ..."
7.2	2.3h "... convert between ordinary and standard index form representations ..."
	2.3p "calculate with standard index form"
7.3	2.3c "... convert a recurring decimal to a fraction"
7.4	2.3q "use surds ... in exact calculations, without a calculator; rationalise a denominator ..."

Route mapping

Exercise	D	C	B	A	A*
A	all				
B			1–2	3–6	
C		1–3	4–8	9	
D				1–25	26–36
E				1–3	4–6
F	all				
G	1–5	6	7–10		
H			all		
I		1–11		12–13	14–15
J				1–13	14
K					all

Answers to diagnostic Check-in test

1 a 0.5   **b** 0.375   **c** 0.32

2 a $\frac{3}{25}$ **b** $\frac{5}{8}$ **c** $\frac{11}{100}$

3 a $\frac{29}{35}$ **b** $\frac{9}{8} = 1\frac{1}{8}$ **c** $\frac{41}{50}$

4 a 8   **b** 4   **c** 25   **d** 125

Answer the following questions **without** using a calculator.

1 Change these fractions to decimals.

   a $\dfrac{1}{2} =$

   b $\dfrac{3}{8} =$

   c $\dfrac{8}{25} =$

2 Change these decimals to fractions.

   a $0.12 =$

   b $0.625 =$

   c $0.11 =$

3 Add these fractions.

   a $\dfrac{2}{5} + \dfrac{3}{7} =$

   b $\dfrac{3}{4} + \dfrac{3}{8} =$

   c $\dfrac{3}{10} + \dfrac{13}{25} =$

4 Find the values of the following.

   a $\sqrt{64} =$

   b $\sqrt[3]{64} =$

   c $5^2 =$

   d $5^3 =$

Incorporating exercises:	7A, 7B, 7C, 7D, 7E	**Key words**	
Homework:	7.1	index	reciprocal
Example:	7.1	(pl: indices)	
		powers	

Learning objective(s)

- use indices

Prior knowledge

Pupils should be familiar with 'square' and 'cube' in both numerical and algebraic terms.

Starter

Ask the class for the answer to 6^2 (36). Explain that we can say that 6^2 is 6 to the power of 2. The 2 is the index.

Ask pupils to use their calculators to answer:

$2^5, 2^4, 2^3, 2^2, 2^1, 2^0, 2^{-1}$
$(32, 16, 8, 4, 2, 1, \frac{1}{2})$

$3^5, 3^4, 3^3, 3^2, 3^1, 3^0, 3^{-1}$
$(243, 81, 27, 9, 3, 1, \frac{1}{3})$

Discuss what is happening.

Main teaching points

When multiplying, add the indices. For example, $3^5 \times 3^2 = 3^{5+2} = 3^7$

When dividing, subtract the indices. For example, $3^5 \div 3^2 = 3^{5-2} = 3^3$

With fractional indices, the denominator is the root and the numerator is the power.

For example, $3^{\frac{2}{5}} = \left(\sqrt[5]{3}\right)^2$

A negative index is a reciprocal. For example, $3^{-2} = \dfrac{1}{3^2}$

Common mistakes

With negative indices, not remembering the reciprocal, but changing the number to a negative.
For example, incorrectly writing, $3^{-2} = -3^2 = -9$.

Plenary

For each of the exercises, discuss with the pupils the most important aspect to *learn*. Pupils could then write the points in their books for future revision.

Incorporating exercises:	7F, 7G, 7H
Homework:	7.2
Example:	7.2

Key words
powers
standard form

Learning objective(s)

- change a number into standard form (and vice versa)
- calculate using numbers in standard form

Prior knowledge

Pupils should be able to multiply and divide by 10, 100 and 1000.
They should know that, for example, $10^5 = 10 \times 10 \times 10 \times 10 \times 10 = 100\ 000$ and be shown that
$10^{-3} = \frac{1}{10} \times \frac{1}{10} \times \frac{1}{10} = \frac{1}{1000} = 0.001$.

Starter

Introduce the lesson with the following quick-fire questions:

- What are 5×10, $5 \times 10 \times 10$, $5 \times 10 \times 10 \times 10$? (50, 500, 5000)
- What are 10^1, 10^2, 10^3? (10, 100, 1000)
- What are 5×10^1, 5×10^2, 5×10^3? (50, 500, 5000)
- What are $5 \div 10$, $5 \div 10 \div 10$, $5 \div 10 \div 10 \div 10$? ($\frac{1}{2}$, $\frac{1}{20}$, $\frac{1}{200}$)
- What are 10^{-1}, 10^{-2}, 10^{-3}? ($\frac{1}{10}$, $\frac{1}{100}$, $\frac{1}{1000}$)
- What are 5×10^{-1}, 5×10^{-2}, 5×10^{-3}? ($\frac{1}{2}$, $\frac{1}{20}$, $\frac{1}{200}$)

Main teaching points

Standard form: $a \times 10^n$.
The a can be as low as 1, and can almost reach 10 (9.99999^r), so $1 \le a < 10$.
The n can be any integer, for example, -7, 0, 1, 47, etc.

Common mistakes

When using a calculator, inputting 3×10^2 as $\boxed{3}\ \boxed{\times}\ \boxed{10}\ \boxed{\text{EXP}}$ (or $\boxed{\text{EE}}$) $\boxed{2}$, instead of $\boxed{3}\ \boxed{\text{EXP}}$ (or $\boxed{\text{EE}}$) $\boxed{2}$.

Plenaries

- Give the pupils a few numbers to change into standard form and vice versa.
- Ask the pupils to use a calculator to work out expressions with powers of 10, writing the answers in standard form.
  For example, $(3 \times 10^4) \times (6 \times 10^{-1})$ ($= 1.8 \times 10^4$)
- Without using a calculator, ask pupils to work out expressions such as:
  $(3 \times 10^2) \times (2 \times 10^6)$, $(3 \times 10^2) \times (5 \times 10^6)$, etc. Include some negative indices.
- Without using a calculator, ask pupils to work out expressions involving division.
  For example, $(6 \times 10^2) \div (2 \times 10^4)$, $(3.6 \times 10^2) \div (4 \times 10^{-2})$, etc.

For the last two bullet points, encourage discussion of the best way to show workings.

Incorporating exercise:	71
Homework:	7.3
Example:	7.3

Key words
rational number   terminating
reciprocal   decimal
recurring
    decimal

Learning objectives

- change a fraction to a decimal, and vice versa, (including fractions which become recurring decimals)
- work out reciprocals of whole numbers and fractions

Prior knowledge

Pupils need to be able to multiply decimals by 10, 100, and 1000, and be able to subtract well. They also need to be able to cancel down fractions.

Starter

Give the class the following quick-fire calculations:

- $0.111\ 111\ \dot{1} \times 10$, $0.266\ 66\dot{6} \times 10$, $0.371\ 371\ \dot{3}\dot{7}\dot{1} \times 1000$

- $1.111111 - 0.111111$, $2.666\dot{6} - 0.266\dot{6}$, $371.371371 - 0.371371$

- Find x, giving your answer as a fraction.
  a $9x = 1$ **b** $9x = 2.4$ **c** $999x = 371$
  $(\frac{1}{9}, \frac{24}{90}$ or $\frac{4}{15}, \frac{371}{999})$

Main teaching points

Writing terminating decimals as fractions – always ask "How many decimal places determines the denominator?". For example, 0.24 has 2 decimal places, the 2 is worth $\frac{2}{10}$, and the 4 is worth $\frac{4}{100}$, so divide 24 by 100, that is: $\frac{24}{100}$ and then cancel down if possible.

With recurring decimals, multiply by the smallest power of 10 to 'line up' the digits, so that when the numbers are subtracted the decimal fraction becomes zero. For example:

- Multiply $0.1111\dot{1}$ by 10 so that $1.11111\dot{1} - 0.11111\dot{1} = 1.000000$
- Multiply $0.2424\dot{2}$ by 100 so that $24.24\dot{2}4\dot{2}4 - 0.24\dot{2}4\dot{2}4 = 24.000000$
- Multiply $0.1363636\dot{3}6$ by 100 so that $13.63\dot{6}3\dot{6} - 0.13\dot{6}3\dot{6} = 13.50000$

Reciprocal means 'one over', so the reciprocal of 4 is $\frac{1}{4}$. To find the reciprocal of a fraction, simply invert the fraction, so the reciprocal of $\frac{3}{4}$ is $\frac{4}{3}$.

Common mistakes

With recurring decimals, the mistakes are to multiply by the wrong power of 10. Also, in the GCSE exam, the question involving changing a recurring decimal to a fraction is often a 'prove' or 'show that' question, and very often pupils do not put down sufficient workings, or clear workings that an examiner can follow.

Plenary

Using a calculator, ask the pupils to convert $\frac{2}{5}$, $\frac{2}{6}$, $\frac{2}{7}$, $\frac{2}{8}$, $\frac{2}{9}$ and $\frac{2}{10}$ into decimals. $(0.4, 0.\dot{3}, 0.\dot{2}8571\dot{4}, 0.25, 0.\dot{2}, 0.2)$
Ask, "How do we find the reciprocal of a number? For example, 6 or $\frac{3}{5}$."
Discuss how easy it is to change terminating decimals into fractions.
Discuss how careful you have to be to line up the relevant numbers when converting recurring decimals to fractions.

Incorporating exercises:	7J, 7K	Key words
Homework:	7.4	rationalise
Example:	7.4	surds

Learning objective(s)

● calculate and manipulate surds

Prior knowledge

Pupils need to know the square numbers, how to find factors, and how to multiply out brackets.

Starter

Ask the class the following quick-fire questions:

● $\sqrt{100}$ = ? $\sqrt{4}$ = ?   (10, 2)

● $\sqrt{100} \times \sqrt{100}$ = ? $\sqrt{4} \times \sqrt{4}$ = ? $\sqrt{7} \times \sqrt{7}$ = ?   (100, 4, 7)

● What are the factors of 10, 12, 20, and 48?
(1, 2, 5, 10; 1, 2, 3, 4, 6, 12; 1, 2, 4, 5, 10, 20; 1, 2, 3, 4, 6, 8, 12, 16, 24, 48)

● Expand and simplify $(x + 3)(x - 4)$ and $(x - y)(x + a)$ = ?
$(x^2 - x - 12, \; x^2 + x(a - y) - ay)$

Main teaching points

Keep referring pupils back to the four rules (+, −, ×, ÷):

$$\sqrt{a} \times \sqrt{b} = \sqrt{ab}, \qquad\qquad C\sqrt{a} \times D\sqrt{b} = CD\sqrt{ab},$$

$$\sqrt{a} \div \sqrt{b} = \sqrt{\frac{a}{b}}, \qquad\qquad C\sqrt{a} \div D\sqrt{b} = \frac{C}{D}\sqrt{\frac{a}{b}}$$

These cover the majority of exam questions with surds.

Also, keep reminding pupils that $\sqrt{y} \times \sqrt{y} = (\sqrt{y})^2 = y$, and when simplifying, they should look to factor out square numbers.

For example, when simplifying $\sqrt{24}$, $\sqrt{24} = \sqrt{2} \times \sqrt{12}$ is not helpful, but $\sqrt{24} = \sqrt{4} \times \sqrt{6}$ is helpful as $\sqrt{4} = 2$.

Common mistakes

Not factoring out square numbers, or, if a square number is spotted, forgetting to square root it when simplifying. For example, incorrectly writing $\sqrt{24} = 4\sqrt{6}$ instead of $2\sqrt{6}$.

Differentiation

Pupils will find the multiplying out of brackets such as $(2 + \sqrt{3})(\sqrt{3} + 1)$ much harder compared to simplifying $2\sqrt{3} + 2 + \sqrt{3}\sqrt{3} + \sqrt{3}$ (which is the brackets already multiplied out). Lots of practice is advised for the pupils who find these hard to do.

Plenary

Once the pupils have packed away and are ready to leave, write one question from each part of the homework on the board, and ask what method should be used to answer each question. You don't need to find the answer – just the method.

Overview

8.1 Quadratic graphs **8.2** Solving equations by the method of intersection	This chapter covers drawing and reading values from quadratic graphs.

Context

As with linear graphs, these more complex graphs have a huge variety of uses in science, economics, architecture, etc.

AQA B references

AO2 Number and algebra: Sequences, functions and graphs

8.1, 8.2   2.6e "generate points and plot graphs of simple quadratic functions, then more general quadratic functions; find approximate solutions of a quadratic equation from the graph of the corresponding quadratic function; find the intersection points of the graphs of a linear and quadratic function, knowing that these are the approximate solutions of the corresponding simultaneous equations representing the linear and quadratic functions"

8.2   2.6f "plot graphs of simple cubic functions, the reciprocal function $y = \frac{1}{x}$ with $x \neq 0$, the exponential function $y = k^x$ for integer values of x and simple positive values of k, the circular functions $y = \sin x$ and $y = \cos x$, ...; recognise the characteristic shapes of all these functions"

Route mapping

Exercise	D	C	B	A	A*
A		all			
B			1–10	11–14	15
C					all

Answers to diagnostic Check-in test

1 6   **2** 22   **3** 29   **4** $\frac{1}{8}$ or 0.125   **5** 4.65, –0.65

1 For the equation $y = x + 3$, find y when $x = 3$.

2 For the equation $y = 3x^2 - 2x + 1$, find y when $x = 3$.

3 For the equation $y = 2x^3 - 3x^2 + 4x - 10$, find y when $x = 3$.

4 For the equation $y = (\frac{1}{2})^x$, find y when $x = 3$.

5 Use the 'completing the square' method to solve $x^2 - 4x - 3 = 0$, giving your answers correct to 2 decimal places.

 Module 3: Number

Incorporating exercises:	8A, 8B
Homework:	8.1
Example:	8.1

Key words

intercept	quadratic
maximum	roots
minimum	vertex

Learning objective(s)

● draw and read values from quadratic graphs

Prior knowledge

Pupils must be able to read and plot coordinates (even if given in table form). They must also be able to substitute into algebraic functions. Pupils should have covered 'completing the square' (Chapter 20, Section 20.5) and be able to do the Check-in test and Homework 8.1, question 2g.

Starter

Ask the pupils to draw in their books a set of axes numbered –5 to +5 on the x-axis and 0 to 30 on the y-axis. Get them to plot these points:

x	−5	−4	−3	−2	−1	0	1	2	3	4	5
y	25	16	9	4	1	0	1	4	9	16	25

Then they should join the points with a smooth curve.
Go around the class pointing out unacceptable curves and ways to improve them. Explain this is the $y = x^2$ curve, and it is one they need to learn and recognise.

Main teaching points

Smooth, accurate, one-line curves are acceptable – *nothing* else will do; be *very* strict on this.

Common mistakes

Poorly plotted and poorly drawn curves – in the GCSE all points must be within $\frac{1}{2}$ small square on the graph paper, and the curve must pass within $\frac{1}{2}$ a small square as well. The curve must be one continuous line, with no gaps or multiple attempts at it.

Plenary

Ask the following questions:
● What shapes are $y = x^2$ and $y = -x^2$ graphs? Where do they intercept with the axes?
● What shapes are $y = 2x^2$, $y = 3x^2$, $y = 5x^2$ graphs? (Sketch them on the same axes.)
● Where do the following graphs intercept with the axes?
$y = x^2 + 2$, $y = 2x^2 + 2$, $y = 3x^2 + 2$

Incorporating exercise:	8G
Homework:	8.2
Example:	8.2

Learning objective(s)

● solve equations by the method of intersecting graphs

Prior knowledge

Pupils should be practised in drawing quadratic graphs. Pupils must also be able to subtract in algebraic terms.

Starter

Ask for the answers to the following:

$$\begin{array}{r} +2 \\ - \; -1 \\ \hline \\ \hline \end{array}$$ (Answer: $+3$)

$$\begin{array}{r} -2x + 2 \\ - \; -3x - 1 \\ \hline \\ \hline \end{array}$$ (Answer: $x + 3$)

$$\begin{array}{r} x^3 - 2x + 2 \\ - \; x^3 - 3x - 1 \\ \hline \\ \hline \end{array}$$ (Answer: $x + 3$)

$$\begin{array}{r} y = x^3 - 2x + 2 \\ - \; 0 = x^3 - 3x - 1 \\ \hline \\ \hline \end{array}$$ (Answer: $y = x + 3$)

Now ask pupils to sketch the graph of $y = x + 3$.

Main teaching points

When doing the algebraic subtractions, do them carefully, taking time to get the signs correct. Draw the graphs carefully to get accurate answers.

Common mistakes

Not clearly showing the method of subtraction in an exam can lead to no marks, as pupils often make mistakes with the signs when doing these.

Plenary

Ask the pupils for the steps that need to be taken for Homework 8.2, parts **a** and **e**.

Variation

9

Overview

9.1 Direct variation
9.2 Inverse variation

This chapter will show how to solve problems where two variables are connected by either a direct relationship or an inverse one. First and foremost, this chapter shows how to work out the multiplication factor (the constant of proportionality, k) between the two variables.

Context

Finding the cost of multiples of goods, the time taken for travel, how many people are needed to do a certain job in a certain time.

AQA B references

AO2 Number and algebra: Calculations

9.1, 9.2   2.3o "calculate an unknown quantity from quantities that vary in direct or inverse proportion"

AO2 Number and algebra: Solving numerical problems

9.1, 9.2   2.4a "draw on their knowledge of operations and inverse operations ... methods of simplification ... in order to select and use suitable strategies and techniques to solve problems and word problems, including those involving ... inverse proportion ..."

AO2 Number and algebra: Equations, formulae and identities

9.1, 9.2   2.5i, "set up and use equations to solve word and other problems involving direct proportion or inverse proportion ..."

Route mapping

Exercise	D	C	B	A	A*
A				all	
B				all	
C				all	

Answers to diagnostic Check-in test

1 a 9   **b** 36   **c** 144   **d** 1   **e** 8   **f** 64   **g** ±10 **h** 4   **i** ±8

2 a 162   **b** $\frac{1}{81}$ or 0.0123456...   **c** $\pm\frac{1}{3}$ or $\pm0.\dot{3}$

3 a 50   **b** 2   **c** 0.02   **d** 0.1

Module 3: Number   © HarperCollins*Publishers* Ltd 2006   77

1 Without using a calculator, write down the answers to the following.

a $3^2 =$

b $6^2 =$

c $12^2 =$

d $1^3 =$

e $2^3 =$

f $4^3 =$

g $\sqrt{100} =$

h $\sqrt[3]{64} =$

i $\sqrt{64} =$

2 Calculate the values of the following expressions if $x = 9$.

a $2x^2 =$

b $\dfrac{1}{x^2}$

c $\dfrac{1}{\sqrt{x}} =$

3 Solve for x, if $y = 10$. Give all your answers to 1 significant figure.

a $x = 5y$

b $y = 5x$

c $x = \dfrac{1}{5y}$

d $y = \dfrac{1}{5x^2}$

Incorporating exercises: 9A, 9B	**Key words**
Homework: 9.1	constant of direct
Example: 9.1	proportionality, proportion
	k direct variation

Learning objective(s)

- solve problems where two variables have a directly proportional relationship
- work out the constant of proportionality

Prior knowledge

Pupils will find that knowing squares, square roots, cubes and cube roots of integers speeds up many of the questions. They must be able to substitute values into algebraic expressions and solve simple algebraic equations.

Starter

Give an aural test on squares, square roots, cubes and cube roots. For example, 3^2, 3^3, $\sqrt{16}$, $\sqrt[3]{8}$, etc. (GCSE requirements are to know the square numbers of 1 to 15, and the cube numbers of 1 to 5.)

$x = 5y$ find x, when $y = 6$ find y, when $x = 15$

$y = 5x^2$ find y, when $x = 3$ find x, when $y = 405$

Main teaching points

Read the question and write what you are told: if A is proportional to B, write $A \propto B$ so $A = kB$, then find k.

Common mistakes

Finding the constant k and then stopping; not completing the question.

Differentiation

Some pupils will be confident with $A \propto B$, but not with $A \propto B^2$ or $A \propto \sqrt{B}$, etc., so will require some extra encouragement and help.

Plenary

Ask the pupils to provide the initial method, if the question starts:
- x is directly proportional to y
- x varies directly to y
- x is directly proportional to y^3
- x varies directly to the square root of y.

			Key words	
Incorporating exercise:	9C		constant of	inverse
Homework:	9.2		proportionality,	proportion
Example:	9.2		k	

Learning objective(s)

- solve problems where two variables have an inversely proportional relationship
- work out the constant of proportionality

Prior knowledge

Pupils will find that knowing squares and square roots, cubes and cube roots of integers speeds up many of the questions. They must be able to substitute values into algebraic expressions and solve simple algebraic equations.

Starter

Pose the following problems:
- If Tom (or pick a pupil from the class) hired a taxi to take him ice-skating, it might cost him £30.
- If Tom and a friend went, how much would it cost them each?
- If Tom and two friends went, how much would it cost them each?
- If Tom and three friends went, how much would it cost them each?
- If Tom and four friends went, how much would it cost them each?

In other words, the more that go, the less it costs each of them – this is inverse proportion.

Main teaching points

Read the question and write out what you are told in the question: if A is inversely proportional to B, write

$A \propto \frac{1}{B}$ so $A = k\frac{1}{B}$ or $A = \frac{k}{B}$, then find k.

This is very similar to direct proportion except the 'inversely' bit makes the second variable a reciprocal.

Common mistakes

Ignoring the 'inversely' and treating it as direct proportion. Incorrect rearranging of the equation, after substituting, to find the value of the required variable.

Plenary

Ask the pupils to provide the initial method if the question starts:
- x is inversely proportional to y
- x varies inversely to y
- x is inversely proportional to y^3
- x varies inversely to the square root of y.

Overview

10.1 Limits of accuracy
10.2 Problems involving limits of accuracy

This chapter deals with how to find the limits of numbers, rounded to various accuracies, and how to use these limits in calculations.

Context

The quality of anything from a biro to a BMW is determined by how accurately the individual parts fit together.

AQA B references

AO2 Number and algebra: Using and applying number and algebra

10.1, 10.2   2.1m "recognise the significance of stating constraints and assumptions when deducing results; . . ."

AO2 Number and algebra: Solving numerical problems

10.1, 10.2   2.4b ". . . recognise limitations on the accuracy of data and measurements"

Route mapping

Exercise	D	C	B	A	A*
A		12			
B			all		
C				1–4	5–12

Answers to diagnostic Check-in test

1 a 13 610   **b** 13 600   **c** 14 000

2 a 14.6   **b** 10   **c** 14.59   **d** 15   **e** 14.592   **f** 14.6

3 22.5625 cm^2 **4** 531.441 cm^3 **5** 16 cm   **6** 600 cm^2

1 Round off 13 605 to:
   a the nearest 10.

   b the nearest 100.

   c the nearest 1000.

2 Round off 14.592 04 to:
   a 1 decimal place   **b** 1 significant figure
   c 2 decimal places   **d** 2 significant figures
   e 3 decimal places   **f** 3 significant figures

3 Find the area of a square of side length 4.75 cm.

4 Find the volume of a cube of side length 8.1 cm.

5 Find the side length of a square with an area of 256 cm^2.

6 Find the surface area of a cube of volume 1000 cm^3.

                                          Module 3: Number

Incorporating exercises:	10A, 10B
Homework:	10.1
Examples:	10.1

Key words

continuous data	lower bound
discrete data	rounding error
limits of accuracy	upper bound

Learning objective(s)

● find the limits of accuracy of numbers that have been rounded to different degrees of accuracy

Prior knowledge

Pupils must be familiar with all types of rounding: to the nearest 1, 10, 100, etc., as well as to a given number of decimal places or significant figures.

Starter

Give pupils a number, for example 6513.9604, and ask them to round off to the nearest 1000, 100, 10, 1, 1 dp, 2 dp, 3 dp, 1 sf, 2 sf, 3 sf, 4 sf, etc.

Main teaching points

A rounded number could have a variety of real values. Get the pupils to look at their answers to the starter questions – all of them are different to the number started with. If a number is rounded to, say, 1 decimal place, the limits will be $\pm\frac{1}{2}$ of 0.1, that is ±0.05.

Common mistakes

Giving an incorrect upper limit. For example, saying that 22 cm to the nearest cm has an upper limit of 22.49 cm (rather than 22.49^r cm, or more usually, 22.5 cm).

Differentiation

For pupils that find this difficult, stick to rounding off to the nearest whole number, 10, 100, etc. as this tends to be easier than the decimal place and significant figure questions.

Plenary

Give the class this problem:
A square tile has a side length of 15 cm to the nearest cm (or 2 sf).
What are the limits of accuracy? (14.5 cm ≤ length < 15.5 cm)
What are the upper and lower bounds? (14.5 cm and 15.5 cm)

Now give them this problem:
10 of these tiles are in a row.
What is the longest possible length? (155 cm)
What is the shortest possible length? (145 cm)

Incorporating exercise:	10C
Homework:	10.2
Examples:	10.2

Key words
limits of
  accuracy
maximum
minimum

Learning objective(s)

● combine limits of two or more variables together to solve problems

Prior knowledge

Pupils need to have successfully completed Exercise 10B.

Starter

The last plenary involved a square tile, side length 15 cm to the nearest centimetre (or 2 significant figures). The pupils worked out that 10 tiles could range in length from 145 cm up to 155 cm. Remind the pupils of this, and extend the question further:

● I need a length of 3 m to be tiled. What is the minimum number of tiles I need to guarantee to cover the 3 m? (21)
● What is the smallest area of 1 tile? (210.25 cm^2)
● What is the largest area of 1 tile? (240.25 cm^2)

Main teaching points

Go over the maximum and minimum table on page 235 of the Pupil Book, using relatively easy values, so the pupils can prove to themselves that each method is true. It will also reinforce for them that there is no substitute for thinking through a problem.

Common mistakes

Using a maximum value rather than a minimum value (or vice versa) in a problem involving subtraction, multiplication or division.

Plenary

Tell the class, "Today I have 50p in my pocket, to the nearest 10p. How much could I actually have?" (45p–55p)
"Yesterday I had five times as much as today. How much could I have had?" (225p–275p)
"I actually have a 50p piece. Its face has an area of 6 cm^2 to the nearest cm^2, and is 1.5 mm thick, to 1 decimal place. How do we work out the lower and upper bounds of the volume of the 50p piece?"

Shape

Overview

11.1 Circumference and area of a circle
11.2 Area of a trapezium
11.3 Sectors
11.4 Volume of a prism
11.5 Cylinders
11.6 Volume of a pyramid
11.7 Cones
11.8 Spheres

This chapter covers the surface area and volume of prisms and solids required at Higher tier, as well as the area and perimeter of circles, sectors and trapezia.

AQA B references

AO3 Shape, space and measures: Geometrical reasoning

11.2, 11.4–11.8
3.2j "... solve problems involving surface areas and volumes of prisms, pyramids, cylinders, cones and spheres; solve problems involving more complex shapes and solids, including segments of circles and frustums of cones"

AO3 Shape, space and measures: Measures and construction

11.1, 11.3
3.4d "... find circumferences of circles and areas enclosed by circles, recalling relevant formulae; calculate the lengths of arcs and the areas of sectors of circles"

Route mapping

Exercise	D	C	B	A	A*
A	1–6	7–12			
B	1–3	4–7			
C				1–4	5–8
D		1–5	6–9		
E			1–10	11	
F			1–2	3–7	8
G			1–5	6–7	
H				all	

Answers to diagnostic Check-in test

1 a 24 cm   b 27 cm^2 2 a 26 cm   b 24 cm^2

3 a 30 cm   b 30 cm^2 4 a 96 cm^3 b 136 cm^2

5 a 28.3 cm   b 63.6 cm^2

1 What is the perimeter and area of this rectangle?

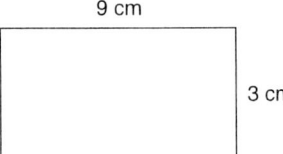

 a Perimeter =_____cm

 b Area =_____cm^2

2 What is the perimeter and area of this parallelogram?

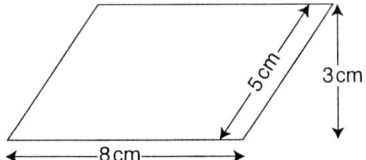

 a Perimeter =_____cm

 b Area =_____cm^2

3 What is the perimeter and area of this triangle?

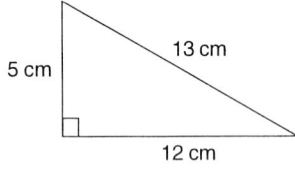

 a Perimeter =_____cm

 b Area =_____cm^2

4 What is the volume and total surface area of this cuboid?

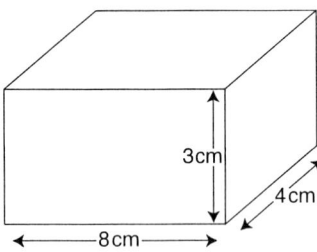

 a Volume =_____cm^3

 b Total surface area =_____cm^2

5 What is the circumference and area of this circle?

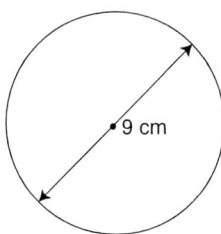

 a Circumference =_____cm

 b Area =_____cm^2

    Module 5: Algebra and Space, shape and measure

Incorporating exercise:	11A	Key words
Homework:	11.1	π
Examples:	11.1	area
		circumference

Learning objective(s)

● calculate the circumference and area of a circle

Prior knowledge

Pupils should have covered the area and circumference of a circle previously. The content of this section is mainly for revision purposes.

Starter

Ask pupils to recall the formulae for the circumference and area of a circle.
Ask them to rearrange the formulae to find expressions for **i** the radius in terms of the circumference, **ii** the radius in terms of the area, and **iii** the area in terms of the circumference.

Main teaching points

For a circle of radius r and diameter d, the circumference is given by $C = 2\pi r$ or $C = \pi d$.

The area of a circle of radius r is given by $A = \pi r^2$. This can be rearranged to give $r = \sqrt{\frac{A}{\pi}}$.

A sketch proof of $A = \pi r^2$ could be offered to pupils. Ask them to imagine a circle cut into slices.

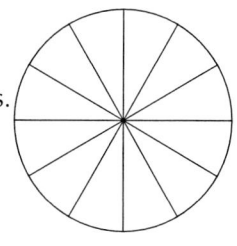

Now rearrange the slices in the form of an approximate parallelogram.

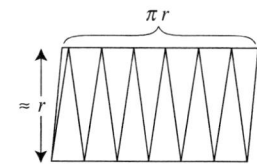

The top and bottom of the parallelogram must equal the circumference of the circle. As the circumference of a circle is given by $C = 2\pi r$, each side must be πr. The length of each triangular section is approximately r. Therefore, the area of a circle is approximately $\pi r \times r = \pi r^2$. The more able pupils should be able to see how the diagrammatical approximation improves as the circle is divided into more slices.

Common mistakes

Less able pupils may be confused between πr^2 and $(\pi r)^2$. They may find it helpful to write out the formula for the area of the circle in the form $A = \pi \times r \times r$ until their confidence in using the formula (and their calculator) improves.

Plenary

The bull's-eye illusion consists of five concentric circles.
Draw the diagram on the board and ask pupils whether they think the shaded inner area is larger than the shaded outer area. Most people tend to think that it is.
Suppose the innermost ring has radius 1cm, the next 2cm and so on.
Ask pupils to work out the two areas. They should find they are both the same area.

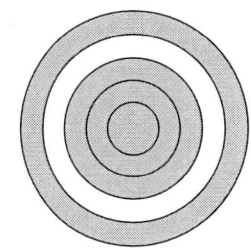

			Key words
Incorporating exercise:	11B		trapezium
Homework:	11.2		
Examples:	11.2		

Learning objective(s)

● find the area of a trapezium

Prior knowledge

Pupils should know how to find the areas of triangles, rectangles and parallelograms.

Starter

Ask pupils what is meant by a 'trapezium'? (It is a quadrilateral with one pair of parallel sides.)
Does a rectangle count as a trapezium? What about a square?

Main teaching points

The area of a trapezium is given by A = half the sum of the parallel sides × height.

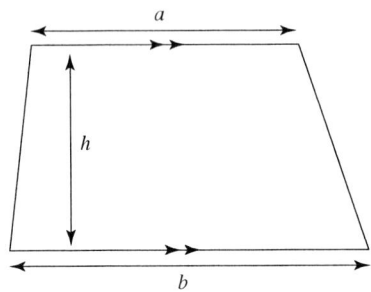

$$A = \frac{1}{2}(a + b)h$$

This can be seen to be the case by splitting a trapezium into two triangles.

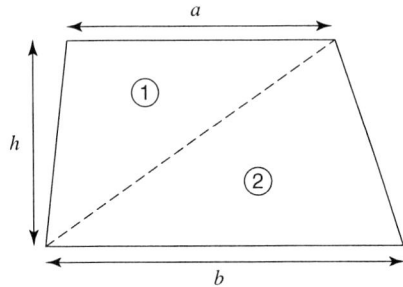

Area of triangle 1 $= \dfrac{1}{2} \times$ base $\times$ height $= \dfrac{1}{2}ah$

Area of triangle 2 $= \dfrac{1}{2} \times$ base $\times$ height $= \dfrac{1}{2}bh$

Therefore, total area $= \dfrac{1}{2}ah + \dfrac{1}{2}bh = \dfrac{1}{2}h(a + b) = \dfrac{1}{2}(a + b)h$

It must always be emphasised in this context that height means perpendicular height, and that we add the two parallel sides. Ensure final answers are given to a suitable degree of accuracy, normally three significant figures.

Common mistakes

The most common error is to calculate $(\frac{1}{2}a + b)$, which is generally down to a lack of understanding in the use of the calculator. This can be avoided by encouraging pupils to label each trapezium with letters a, b and h before commencing the question and by instructing them to set out their working appropriately as shown in the worked examples.

Plenary

Draw the following diagrams on the board:

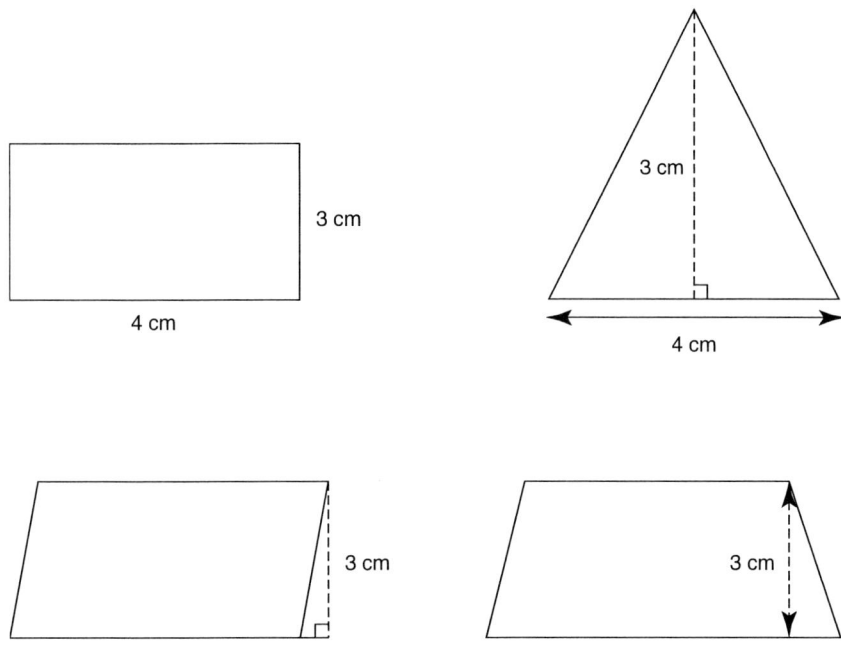

Which has the largest area?

Incorporating exercise:	11C	Key words
Homework:	11.3	area
Examples:	11.3	sector

Learning objective(s)

● calculate the length of an arc and the area of a sector

Prior knowledge

Pupils need to be able to calculate the area and circumference of a circle, as covered in Section 11.1 of the Pupil Book.

Starter

Following their work on the area and circumference of circles in 11.1, most pupils should be in a position to begin to discover the formulae for the arc perimeter and area of a sector.
Ask pupils to write down the formula for the circumference of a circle. Then ask them to write a formula for the curved part of a semi-circle, and again for a quarter circle. Can they make the leap to finding the formula for an arc sector, given a sector angle? Assist pupils if necessary by, for example, drawing a sector with a given sector angle and asking what fraction of the circle (or circumference) has been drawn here?
The area of a sector could be discovered in a similar fashion.

Main teaching points

If an arc AB subtends an angle θ° at the centre of a circle then:

the arc length AB is given by Arc length $= \dfrac{\theta}{360} \times 2\pi r$,

the area of the sector is given by Area $= \dfrac{\theta}{360} \times \pi r^2$.

Plenary

Ask pupils to rearrange the formulae for arc length and the area of a sector so that they can derive simplified expressions for **i** the radius in terms of the arc length and **ii** the radius in terms of the area.

Incorporating exercise:	11D	Key words
Homework:	11.4	cross-section
Examples:	11.4	prism

Learning objective(s)

- calculate the volume of a prism

Prior knowledge

Pupils need to be able to calculate areas of rectangles, triangles, trapezia and compound shapes.

Starter

Draw a rectangle, triangle, circle and trapezium on the board. Ask pupils to recall the areas of these shapes. Extend these diagrams into 3-D prisms by giving them depth. Label them: rectangular prism, triangular prism, circular prism, trapezoidal prism.
Ask pupils what the rectangular prisms and circular prisms are more commonly called. Why is it that the other two prisms do not have special names?

Main teaching points

The volume of a prism is given by the product of the area of the cross-section of the prism and its length. This might more usefully be written as $V = A \times$ length.

For most problems, it is therefore important to stress that pupils calculate the cross-sectional area of the prism as an essential first step.

Common mistakes

Pupils will sometimes mistake the length of the prism for a length occurring in the common cross-section. They should be able to avoid this mistake by drawing a good diagram and solving the problem in two distinct stages.

Plenary

Draw a diagram of a cylinder on the board, labelling the radius and height. By considering the formula $V = CSA \times$ length, ask pupils to find the formula for the volume of a cylinder. Ask them to use this formula to find the volume of a cylinder with radius 2 cm and height 3 cm.

Incorporating exercise:	11E	Key words
Homework:	11.5	cylinder
Examples:	11.5	surface area
		volume

Learning objective(s)

● calculate the volume and surface area of a cylinder

Prior knowledge

Pupils should know how to find the volume of a prism, as covered in Section 11.4, and be able to calculate the area of a circle, as covered in Section 11.1.

Starter

Revise or remind pupils of calculations involving areas of circles. Ask them, for example, what the area of a circle of diameter 3 cm is. If a circle's area is 25 cm², what is its radius?

Main teaching points

A cylinder is a circular prism with a circular cross-section. The volume of this prism is therefore $V = \text{CSA} \times \text{length} = \pi r^2 \times h$. So for a cylinder of radius r and height h, $V = \pi r^2 h$.

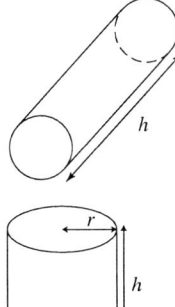

Consider a piece of paper wrapped round a cylinder of radius r and height h. This could be unwrapped to give a rectangular piece of paper, of dimensions $2\pi r$ by h. The curved surface area of a cylinder is therefore $2\pi rh$.

There are also two circles at the top and bottom of the cylinder each having area πr^2.
The total surface area of a cylinder is therefore $2\pi r^2 + 2\pi rh = 2\pi r(r + h)$·

The formula for the volume of a cylinder can be rearranged to find the radius or the height. So:

$$h = \frac{V}{\pi r^2} \text{ and } r = \sqrt{\frac{V}{\pi h}} \cdot$$

Inexact answers are usually given to three significant figures.

Common mistakes

As in Section 11.1, less able pupils may be confused between πr^2 and $(\pi r)^2$. They may find it helpful to use a simpler version of the formula in the form $V = \pi \times r \times r \times h$ until their confidence improves.

Plenary

A cylinder has radius r and height h. A second cylinder has radius $2r$ and height $\frac{h}{2}$. Which cylinder has the largest volume and by how much?

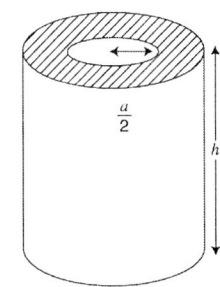

A solid cylinder of radius a and height h has a smaller cylinder of radius $\frac{a}{2}$, height h drilled into it and the material removed.
What is the volume of the remaining shape? What is the total surface area of the remaining shape?

Incorporating exercise:	11F	Key words	
Homework:	11.6	apex	pyramid
Examples:	11.6	frustum	volume

Learning objective(s)

● calculate the volume of a pyramid

Prior knowledge

Pupils should know how to calculate the areas of simple shapes such as circles, rectangles and triangles. They should have ideally completed the work on finding the volume of a prism in Section 11.5. Knowledge of Pythagoras' theorem is required for some of the harder problems (such as Homework question 6).

Starter

Define a pyramid, then ask pupils to draw diagrams of various pyramids.
Draw and label a square-based pyramid, a triangular-based pyramid, a hexagonal-based pyramid and a circular-based pyramid.
Ask pupils for the common name for a circular-based pyramid.

Main teaching points

A pyramid is a 3-D shape with a base which rises to a common vertex, called the apex. A frustum is formed by cutting off the top of a pyramid.

Pyramids are typically illustrated as right-pyramids – pyramids where the apex is vertically above the centre of mass of the pyramid – but other pyramids should be expected to be seen in the Higher tier at GCSE. There are examples of pyramids that are not right-pyramids in Exercise 11F in the Pupil Book: see questions 1c and 1d.

For a pyramid on a base of any shape, volume $= \frac{1}{3} \times$ area of base $\times$ perpendicular height. That is $V = \frac{1}{3}Ah$.

The proof is beyond the scope for GCSE, but is given here for completion. Consider a pyramid of area A at its base. As the lengths of the lines decrease linearly towards the vertex, the area therefore decreases quadratically. The area of the cross-sectional slice at height z above the base is therefore given by:

$$A(z) = A\frac{(h-z)^2}{h^2}$$

The volume is therefore:

$$V = \int_0^h A(z)\,dz = A\int_0^h \frac{(h-z)^2}{h^2}\,dz = A\int_0^h \frac{h^2-2hz+z^2}{h^2}dz = A\int_0^h \left(1-2h^{-1}z+h^{-2}z^2\right)dz$$

$$= A\left[z - h^{-1}z^2 + \frac{1}{3}h^{-2}z^3\right]_0^h = \frac{1}{3}Ah$$

Plenary

Ask pupils to find the volume of a circular-based pyramid (that is, a cone) of radius r, height h.
Ask pupils to find the volume of this frustum.

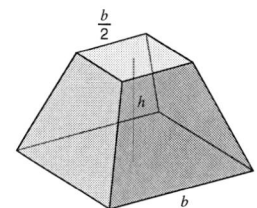

Incorporating exercise: 11G	**Key words**
Homework: 11.7	slant height vertical height
Examples: 11.7	surface area volume

Learning objective(s)

● calculate the volume and surface area of a cone

Prior knowledge

Pupils should know how to find the volume of a pyramid, as covered in Section 11.6 of the Pupil Book. Some problems also require the use of Pythagoras' theorem.

Starter

Recall the formula for the volume of a cone from the plenary in 11.6, and ask pupils to calculate the volume of a cone of base radius 5 cm and height 10 cm.
Ask pupils how the volume changes if the height of this cone is doubled? How does the volume change if the radius is doubled? Will this work for any cone? Why?

Main teaching points

A cone is a pyramid on a circular base: r is the base radius, h is the perpendicular height and l is the slant height.

The volume of a cone is derived from the formula for the volume of a pyramid.

That is, $\frac{1}{3}$ × area of base × perpendicular height, so:

$$V = \frac{1}{3} \times \pi r^2 \times h \qquad V = \frac{1}{3}\pi r^2 h$$

The curved surface area S is given by $S = \pi r l$, which is easy to prove from the work on sectors covered in Section 11.3. The cone can be 'opened up' to form a sector of a circle as shown on the right:

The complete circle would have had circumference $2\pi l$ and so the area of the sector is given by $\frac{2\pi r}{2\pi l} \times \pi l^2 = \pi r l \cdot$

The total surface area of the cone is therefore given to be:
$A = \pi r l + \pi r^2 = \pi r(l + r)$

A frustum of a cone is obtained by removing a smaller cone from the top of a larger cone.

Pupils should be prepared to leave their answers in terms of π if required to do so.

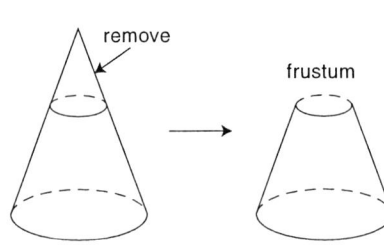

Plenary

A frustum is formed by removing a cone of height h from a cone of height $2h$. Find the volume of the frustum.

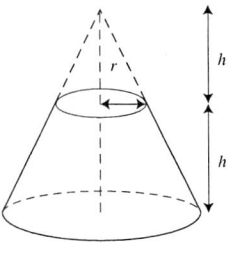

Incorporating exercise:	11H	Key words
Homework:	11.8	sphere
Examples:	11.8	surface area
		volume

Learning objective(s)

⬤ calculate the volume and surface area of a sphere

Prior knowledge

Pupils should have covered Sections 11.1–11.7.

Starter

At this stage, and at the end of a long chapter, it is as well to consolidate the material covered so far. Draw a table with prism, cylinder, pyramid, cone, sphere along one row. Label the next row volume and the following row total surface area. Ask the pupils to fill in or suggest formulae for each empty box, omitting the surface area formulae for the pyramid and prism.
The formulae for the sphere might already be known by some pupils, and could then be discussed.

Main teaching points

The volume of a sphere of radius r is given by $V = \dfrac{4}{3}\pi r^3$.

The (curved) surface area of a sphere is given by $A = 4\pi r^2$.

To find the radius of a sphere, $V = \dfrac{4}{3}\pi r^3$ can be rearranged to give $r = \sqrt[3]{\dfrac{3V}{4\pi}}$.

Inexact answers are usually given to three significant figures.

Common mistakes

There may still be mistakes due to the misinterpretation of $\dfrac{4}{3}\pi r^3$ as $\left(\dfrac{4}{3}\pi r\right)^3$. It is useful to point out this potential error when going through at least one example of finding the volume of a sphere.

Plenary

Pose this problem:
Suppose 1000 ball bearings, each of radius 0.5 cm, are placed into a cubical tank having a length of 10 cm. What percentage of the tank is filled with ball bearings?

Pythagoras and trigonometry

Overview

12.1 Pythagoras' theorem
12.2 Finding a shorter side
12.3 Solving problems using Pythagoras' theorem
12.4 Trigonometric ratios
12.5 Calculating angles
12.6 Using the sine function
12.7 Using the cosine function
12.8 Using the tangent function
12.9 Which ratio to use
12.10 Solving problems using trigonometry

This chapter covers Pythagoras' theorem and the trigonometry of right-angled triangles with their associated applications.

Context

Pythagoras' theorem has many real-life applications. It has long been a recognised method for constructing accurate right-angles, for example in building walls which need to be at right-angles to each other, or calculating the length of restraining wires or stays of a mast or pole.

The cosine rule can be presented as a generalised form of Pythagoras' theorem, and this along with the sine rule, opens up a wide range of practical problems that may then be solved, from bearing problems to situations involving angles of elevation and depression.

AQA B references

AO3 Shape, space and measures: Geometrical reasoning

12.1–12.3   3.2g "understand, recall and use Pythagoras' theorem in 2-D, then 3-D problems;..."
12.4–12.10   3.2h "...understand, recall and use trigonometrical relationships in right-angled triangles, and use these to solve problems, including those involving bearings..."

Route mapping

Exercise	D	C	B	A	A*
A		all			
B		all			
C		all			
D			all		
E			1–5	6–7	
F			all		
G			all		
H			all		
I			all		
J			all		
K			1–3	4	
L			1–9	10	
M			all		
N			1–7	8	
P			all		

Answers to diagnostic Check-in test

1 1, 4, 9, 16, 25, 36, 49, 64, 81, 100

2 a 97 **b** 135 **c** 324

3 a 4.8 **b** 18.0 **c** 21.0

4 a 2.68 **b** 12.7 **c** 0.155

5 a 5.48 **b** 16.9 **c** 0.762

6 a Correct drawing, with line north 3 cm and line west 4 cm
  b 10 km

1 Write down the first **ten** square numbers.

2 Work out:

   a $4^2 + 9^2$ **b** $12^2 - 3^2$ **c** 9×6^2

3 Round these numbers to one decimal place.

   a 4.793   **b** 18.0165   **c** 20.96

4 Round these numbers to three significant figures.

   a 2.684   **b** 12.678   **c** 0.1552

5 Work out these square roots on your calculator, giving your answers correct to three significant figures.

   a $\sqrt{30}$ **b** $\sqrt{285}$ **c** $\sqrt{0.58}$

6 a Using a scale of 1 cm to represent 2 km, draw a scale diagram of the journey of a boat which sails 6 km north and then 8 km west.

   b By measuring your scale drawing, find the distance in a direct line back from the boat's finishing point to its starting point.

 Module 5: Algebra and Space, shape and measure

Incorporating exercise:	12A
Homework:	12.1
Examples:	12.1

Key words
hypotenuse
Pythagoras'
theorem

Learning objective(s)

● calculate the length of the hypotenuse in a right-angled triangle

Prior knowledge

Pupils must know how to find the square and square root of a number. It is also helpful if pupils have a familiarity with square numbers.

Pupils must also know how to round numbers correctly, and should have an appreciation of what is a suitable level of accuracy given the context of the problem involved.

Starter

Display a target board of whole numbers and ask pupils to find the square numbers. Ask for any known square numbers which are missing from the board. Try to elicit the first 12 square numbers, with some other commonly known ones such as 400, 625, 1 000 000.
Ask for a definition of square numbers. Ask for squares of some simple decimal numbers such as 0.5 and 0.1.

Main teaching points

Pupils should understand that Pythagoras' theorem equates the area of a square drawn on the hypotenuse to the sum of the two squares drawn on the two shorter sides of a right-angled triangle. It should be made clear to them that the benefit of this theorem is that it allows a method to be developed for finding an unknown length. For the purposes of this exercise, all that is to be found is the length of the hypotenuse.

Pupils should understand that this theorem only holds for right-angled triangles, and that it is incorrect to try to apply it in other situations. Therefore, unless they know positively that a triangle is right-angled, this theorem should not be applied.

It should be made clear to pupils how to identify correctly the hypotenuse of a right-angled triangle, that it will always be the side facing the right angle.

A formal method for finding the length of the hypotenuse should now be introduced, following the layout demonstrated in the Pupil Book. Pupils should be instructed to follow the method clearly and precisely.

Common mistakes

Perhaps the most common mistake is forgetting to take the square root to find the final answer. This can generally be avoided if sufficient attention is paid to the layout of solutions. The other common mistake is that pupils often try to apply the theorem in unsuitable situations, such as when the triangle in question is not known to be a right-angled triangle.

Differentiation

More able pupils will finish this work quite quickly, and if time allows, they can be asked to try and find other examples of Pythagorean triples, such as those in questions 7, 8 and 9.

Plenary

Ask for a summary of key points, including how to identify the hypotenuse, an algebraic statement of the theorem and a brief summary of the method.
If some pupils have had time to look for triples, ask them to feed back to the group.

Incorporating exercise:	12B
Homework:	12.2
Examples:	12.2

Key words
Pythagoras'
theorem

Learning objective(s)

● calculate the length of a shorter side in a right-angled triangle

Prior knowledge

As in Section 12.1, pupils must know how to find the square and square root of a number. It is also helpful if pupils have a familiarity with square numbers.

Pupils must know how to round numbers correctly, and should have an appreciation of what is a suitable level of accuracy given the context of the problem involved.

For this section, pupils must be familiar with using Pythagoras' theorem to find the length of the hypotenuse of a right-angled triangle.

Starter

Give the class the number 100 as a target, and using a target board or a list of numbers, ask what needs to be added to each to give the target number. Repeat for other target numbers, such as 64, 81, 25. Ask what the target numbers have in common, trying to elicit that they are all square numbers. Repeat the task using targets such as 7^2 or 12^2. Ask what type of calculation has to be done to find the required number, aiming to elicit subtraction.

Main teaching points

Pupils should understand that, from the original statement of Pythagoras' theorem, a method for calculating the length of a missing side can be obtained by rearranging the sum into a difference. It must be made very clear that, in order to avoid possible errors, the hypotenuse must be clearly identified each time. Pupils should be reminded that they must add to find the hypotenuse (longest side), and subtract to find a shorter side.

Pupils should be reminded that all uses of Pythagoras' theorem only apply to right-angled triangles, and they may need to be reminded about suitable levels of accuracy.

Common mistakes

Pupils often confuse the two methods seen in Sections 12.1 and 12.2, and this is usually due to a lack of care when identifying the hypotenuse. The common mistakes highlighted in Section 12.1 still apply in this section as well.

Differentiation

Questions 2 and 3 of Exercise 12B in the Pupil Book involve cases of both types encountered so far. Some require finding the hypotenuse and others require finding one of the shorter sides. Lower achieving pupils will need this pointing out, and will need to be reminded of the difference between the two methods.

Plenary

Ask pupils for a summary of the two principle methods, and when to use them.
Ask pupils how to identify the hypotenuse of a right-angled triangle.

Incorporating exercise:	12C, 12D, 12E	Key words
Homework:	12.3a, 12.3b, 12.3c	3-D Pythagoras'
Examples:	12.3a, 12.3b, 12.3c	isoceles triangle theorem

Learning objective(s)

- solve problems using Pythagoras' theorem

Prior knowledge

Pupils must know how to use Pythagoras' theorem to find either the hypotenuse or a shorter side.
This section provides a wide variety of situations in which problems can be solved using Pythagoras' theorem. The extra skills and knowledge required include:
- use of compass directions
- use of coordinates
- finding the area of a rectangle
- the properties of isosceles triangles.

Starter

Ask pupils to suggest places around the classroom that have right angles.
Show pupils a coordinate grid with some points plotted. Select two of the points and ask pupils how they might find the distance between these points. Find the actual distance using Pythagoras' theorem and repeat the process for another couple of points. Can they suggest a general formula for the distance between two coordinates (x_1, y_1) and (x_2, y_2)?

Main teaching points

Pupils need to be able to identify a right-angled triangle in a wide variety of situations, and to understand how Pythagoras' theorem can be applied to solve problems in context.

Pupils should be encouraged to start by drawing a diagram of the right-angled triangle involved in each situation, clearly labelling the right angle, known lengths and the unknown side with an x.

Pupils will benefit from a demonstration of finding a right-angled triangle in as many different situations as possible. In particular, they will need a demonstration of the type of situation which involves a journey given in compass directions, and of the type involving finding the distance between two coordinate points.

There are two additional applications in this section that are unique to the Higher syllabus. The first is the application of Pythagoras' theorem to isosceles triangles. These examples usually involve bisecting the triangle so that a perpendicular height may be found using Pythagoras' theorem. Once this is known, the area of the triangle can then easily be calculated. This is an important application and should be demonstrated to pupils.

The second involves applications in three dimensions. Perhaps the most useful example that might be demonstrated is that of finding a diagonal length in a cube or cuboid. At each stage, pupils should be urged to draw out the separate plane triangle that includes the information they are using and the side they are trying to find.

For completeness, pupils should be shown the converse of Pythagoras' theorem: that if $a^2 + b^2 = c^2$ for sides a, b and c of a triangle, then the triangle must contain a right angle.

Common mistakes

Identifying the position of the right angle can cause difficulties for less able pupils. Pupils will often be reluctant to set out their work in stages and so may need some encouragement to do so, particularly in regard to the later questions set in three dimensions.

Differentiation

Lower achieving pupils will have particular difficulty in seeing how to extract the relevant information from the context. These pupils will need more assistance in constructing the diagram which forms the beginning of the solution. This will become particularly apparent in those questions requiring more than one stage of calculation, for example, the applications of Pythagoras' theorem in three dimensions where the ability to visualise planes within three dimensions can pose problems.

Plenary

Ask pupils to summarise the method for using Pythagoras' theorem. Ensure that all problems start with drawing a diagram, and identifying whether the hypotenuse or a shorter side is to be found. How can the hypotenuse be identified? Ask for a reminder of how the method changes, depending on which side is to be found.
A cube has sides of length a. What is the length of one of its diagonals?
A possible generator of Pythagorean triples is given by sides having lengths $m^2 - n^2$, $2mn$, $m^2 + n^2$. What do we mean by a generator, and why do sides having these lengths always produce Pythagorean triples?

			Key words	
Incorporating exercise:	12F		adjacent side	sine
Homework:	12.4		cosine	tangent
Example:	12.4		hypotenuse	trigonometry
			opposite side	

Learning objective(s)

● use the three trigonometric ratios

Prior knowledge

Pupils should be able to round numbers to a given number of significant figures.

Starter

Ask pupils to draw three progressively larger right-angled triangles having the same angle of 30°.

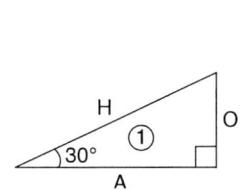 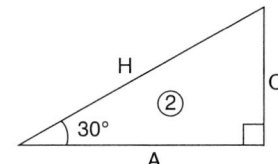 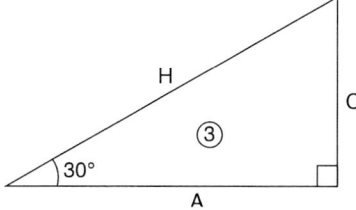

Ask them to measure the lengths of each side as accurately as possible, and to complete this table.

Triangle	Angle	Opposite	Adjacent	Hypotenuse	$\dfrac{O}{H}$	$\dfrac{A}{H}$	$\dfrac{O}{A}$
1	30°						
2	30°						
3	30°						

They should, of course, find the final three columns are (respectively) the same for each angle. Is this true for all angles? Repeat the starter for another angle.

The idea of sine, cosine and tangent can then be introduced and pupils' final three column results can be checked using a calculator.

Main teaching points

The three sides of a right-angled triangle can be labelled as the hypotenuse, opposite and adjacent. The hypotenuse is always the longest side in the triangle and is found opposite the right angle. The side opposite the given angle θ is called the opposite. The remaining side is the adjacent side.

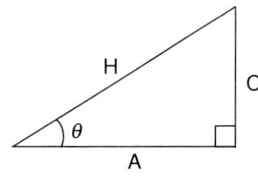

Three ratios can be defined for any given angle θ. These are:

$$\sin\theta = \frac{O}{H}, \quad \cos\theta = \frac{A}{H}, \quad \tan\theta = \frac{O}{A}$$

There are several ways of remembering the trigonometric ratios. Perhaps the easiest and most convenient is SOHCAHTOA, but many others include:

- *Some Of Her Children Are Having Trouble Over Algebra*
- *Tommy On A Ship Of His Caught A Herring*
- *She Offered Her Cat A Heaping Teaspoon Of Acid*

Perhaps pupils could attempt to come up with some of their own?

Confidence in the use of the calculator is crucial. Ensure all pupils have their calculators set to degree mode and are able to follow the correct sequence of key strokes for their calculator when working out, for example, 12 sin 60. More detail is offered in the worked example sheet.

Common mistakes

Some pupils persist in writing sin 30, for example, as 30 sin. This can occur more often with those using the non-DAL calculators as it is the correct key-stroke sequence for their calculator. Some pupils also annoyingly persist in pronouncing sine as "sin".

Plenary

Ask pupils to consider the sine and cosine of complementary angles. For example, ask them to calculate:
- sin 25 and cos 65
- sin 80 and cos 10
- sin 37 and cos 53

In each case they are the same. Why is this? Can pupils come up with a good explanation, or perhaps a formal proof?

Incorporating exercise:	12G	Key words
Homework:	12.5	inverse functions
Example:	12.5	

Learning objective(s)

- use the trigonometric ratios to calculate an angle

Prior knowledge

Pupils should have covered Section 12.4. They should therefore be able to calculate values of trigonometric ratios to a given degree of accuracy.

Starter

Ask pupils to attempt to find angles (to one decimal place) that have respectively a sine of 0.1, 0.2 and 0.3. They are likely to do this through a trial and error process.

Repeat the exercise for cosine. This can then lead on to the discussion of inverse functions, and in particular, the use of the inverse sine and inverse cosine functions on the calculator.

Main teaching points

The inverse sine function on a calculator is used to find the angle whose sine is a given value. Similarly for the inverse cosine and inverse tangent.

Ensure all pupils have their calculators set to degree mode and are able to follow the correct sequence of key strokes for their calculator when working out, for example, $\sin^{-1} 0.5$. This can differ when using DAL calculators, such as the Sharp RL-531VH, compared to other scientific calculators. Two examples are given in Worked example 6.5.

Common mistakes

Pupils often write the meaningless $0.5 \sin^{-1}$ rather than the correct $\sin^{-1} 0.5$. This is most prevalent amongst those using non-DAL calculators, where $0.5 \sin^{-1}$ is the correct sequence of key strokes to find the inverse sine of 0.5.

Plenary

Ask pupils why a calculator returns an error message when an attempt is made to evaluate the inverse sine of a number greater than 1. (Draw several right-angled triangles on the board to try and elicit a useful, if not correct, response.)

Similarly, why is an error message returned when we try to evaluate the inverse cosine of a number greater than 1?

Conversely, why *is* it possible to find the inverse tangent of any value greater than zero?

Incorporating exercise:	12H	**Key words**
Homework:	12.6	sine
Examples:	12.6	

Learning objective(s)

● use the sine function to find lengths of sides and angles in right-angled triangles

Prior knowledge

Pupils should have covered Sections 12.4 and 12.5 and be familiar with the sine function. They should be able to evaluate simple expressions involving the sine and inverse sine functions using their calculator.

Starter

Draw a right-angled triangle on the board, labelling one of the angles θ. Ask pupils how the sides of the triangle should be labelled. Ask them to recall what is meant by the sine of an angle.
Now place numerical values on the opposite side and hypotenuse of the diagram. For example, label the opposite side as being 3 cm and the hypotenuse as 5 cm. What is the value of sin θ in this case?
Practise rearranging simple equations of the type $ab = c$. For example, if $3x = 15$, can pupils find x? Can pupils solve equations such as the following?

$$12x = 5$$
$$10 = \frac{15}{x}$$

Main teaching points

In a right-angled triangle, the sine of an angle θ is given by:

$\sin \theta = \dfrac{O}{H}$, where O is the side of the triangle opposite the angle and H is the hypotenuse.

There are three main types of question in this section:

● finding the opposite side of a right-angled triangle given the hypotenuse
● finding the hypotenuse given the opposite side
● finding the angle given the opposite side and hypotenuse.

Finding a missing side involves rearranging a simple equation, which is where most of the difficulties are likely to lie in both this and the following two sections. Go through plenty of examples with pupils to ensure they gain confidence in this important area.

Differentiation

Less able pupils may still need assistance in identifying the opposite and hypotenuse sides of the triangle. Make sure they get into the habit of labelling the triangle O, A and H before they begin to attempt the question.

Some pupils will still find difficulties in rearranging the equation to find an unknown side. They may be more comfortable with an alternative approach such as the triangle method. For example, if they need to find x from the equation

$$\sin 40 = \frac{5}{x}$$

they can put each term into a triangle like this:

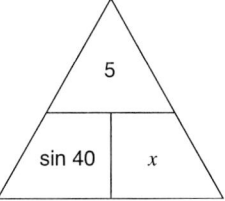

By covering up the unknown quantity required, they can find the correct formula, that is

$$x = \frac{5}{\sin 40} \, .$$

Plenary

The angle 30° is one of the few values of θ for which $\sin \theta$ gives an exact value. Can any pupil explain why $\sin \theta = \frac{1}{2}$? Offer a hint (if needed) by drawing an equilateral triangle on the board and dropping a perpendicular line from one vertex to the midpoint of the opposite side.

Incorporating exercise:	12I	**Key words**
Homework:	12.7	cosine
Examples:	12.7	

Learning objective(s)

• use the cosine function to find lengths of sides and angles in right-angled triangles

Prior knowledge

Pupils should have covered Sections 12.4 and 12.5 and be familiar with the cosine function. They should be able to evaluate simple expressions involving the cosine and inverse cosine functions using their calculator.

Starter

Draw a right-angled triangle on the board, labelling one of the angles θ. Ask pupils how the sides of the triangle should be labelled. Ask them to recall what is meant by the cosine of an angle.
Now place numerical values on the opposite side and hypotenuse of the diagram. For example, label the adjacent side as being 4 cm and the hypotenuse as 5 cm. What is the value of $\cos \theta$ in this case?
Repeat, but this time label the opposite side and the hypotenuse. For example, let the opposite side be 12 cm and the hypotenuse 13 cm. Is it possible to find $\cos \theta$ in this case without first finding θ?

Main teaching points

In a right-angled triangle, the cosine of an angle θ is given by:

$\cos \theta = \dfrac{A}{H}$, where A is the side of the triangle adjacent to the angle and H is the hypotenuse.

There are three main types of question in this section:

• finding the adjacent side of a right-angled triangle given the hypotenuse
• finding the hypotenuse given the adjacent side
• finding the angle given the adjacent side and hypotenuse.

Finding a missing side involves rearranging a simple equation, which is again where most of the difficulties are likely to lie. Go through plenty of examples with pupils to ensure they gain confidence in this area.

Differentiation

Lower ability pupils may need assistance in identifying the adjacent and hypotenuse sides of the triangle. Make sure they get into the habit of labelling the triangle O, A and H before they begin to attempt the question.

As in Section 12.6, some pupils will still have difficulties in rearranging the equation to find an unknown side. Again, they may be more comfortable with the triangle method (as demonstrated in Section 12.6).

Plenary

Investigate the relationship $\cos \theta = \sin(90 - \theta)$ for several values of θ. Can pupils explain why this identity is always true (for angles up to 90°)? If a hint is needed, draw a right-angled triangle on the board, labelling one angle as a right angle and another as θ. What must the size of the third angle be?

Incorporating exercise:	12J	Key words
Homework:	12.8	tangent
Examples:	12.8	

Learning objective(s)

● use the tangent function to find lengths of sides and angles in right-angled triangles

Prior knowledge

Pupils should have covered Sections 12.4 and 12.5 and be familiar with the tangent function. They should be able to evaluate simple expressions involving the tangent and inverse tangent functions using their calculator.

Starter

Draw a right-angled triangle on the board, labelling one of the angles θ. Ask pupils how the sides of the triangle should be labelled. Ask pupils to recall what is meant by the tangent of an angle.
Now place numerical values on the opposite side and adjacent side of the diagram. For example, label the opposite side as being 5 cm and the adjacent as 2 cm. What is the value of $\tan\theta$ in this case?
Repeat, but this time label the adjacent side and the hypotenuse. For example, let the adjacent side be 7 cm and the hypotenuse 25 cm. Is it possible to find $\tan\theta$ in this instance without first finding θ?

Main teaching points

In a right-angled triangle, the tangent of an angle θ is given by

$\tan\theta = \dfrac{O}{A}$ where O is the side opposite to the angle and A is the side adjacent to the angle.

There are three main types of question in this section:
● finding the opposite side of a right-angled triangle given the adjacent side
● finding the adjacent side given the opposite side
● finding an angle given the opposite side and adjacent side

Finding a missing side involves rearranging a simple equation, which is where most difficulties are likely to lie in a similar manner to that in Sections 12.6 and 12.7. Once again, go through several examples with pupils ensuring they are gaining competence in this area.

Differentiation

Lower ability pupils may again need assistance in identifying the opposite and adjacent sides of the triangle. Always ask them to label the triangle O, A and H before they begin to attempt the question.
Again some pupils will have difficulty in rearranging their equation to find an unknown side. They may wish to continue to use an alternative approach such as the triangle method (in a similar fashion to Sections 12.6 and 12.7).

Plenary

Investigate, for several values of θ, the relationship $\tan\theta = \dfrac{\sin\theta}{\cos\theta}$.

Can pupils explain why this identity is always true (for angles up to 90°)? If some direction is needed, draw a right-angled triangle on the board, labelling one angle as θ and the three sides as O, A and H. Ask them to find expressions for:

$\sin\theta$, $\cos\theta$ and $\dfrac{\sin\theta}{\cos\theta}$

Incorporating exercise:	12K		**Key words**
Homework:	12.9		cosine
Examples:	12.9		sine
			tangent

Learning objective(s)

● decide which trigonometric ratio to use in a right-angled triangle

Prior knowledge

Pupils should have covered Sections 12.4–12.8 and be familiar with calculating missing sides and angles of right-angled triangles using the trigonometric ratios sine, cosine and tangent.

Starter

Ask pupils for the definitions of sine, cosine and tangent with regard to right-angled triangles. How do they remember the definitions? Do they use SOHCAHTOA or a mnemonic?

Main teaching points

This section is basically an amalgamation of Sections 12.6–12.8, with the crucial difference that pupils have to decide which trigonometric ratio to use.

To start with at least, ensure pupils correctly label the sides of the triangle as O, A and H (opposite, adjacent and hypotenuse). When attempting to find a missing length, they should then be thinking along the lines of "Which side do I know and which do I need to find?" For example, if they know the opposite and they need to find the adjacent, they will need to use tan.

Finding a missing angle involves a similar procedure. Once the sides of the triangle are correctly labelled, pupils should be thinking "I know the opposite and hypotenuse, so I will be using... sine."

Emphasise that they must write down the formula they intend to use, so that it contains both known and unknown quantities.

Differentiation

Less able pupils will always need to label the triangle O, A and H before they begin to attempt the question. All pupils should endeavour to do this to start with, although it will soon become superfluous for the more able.

Some less able pupils will do their utmost to avoid putting down even the minimum amount of working necessary. Emphasise to them that an initial equation must be seen, so that (method) marks can then be given even if their final answer is wrong or incorrectly rounded.

Plenary

Demonstrate to pupils the consistency of Pythagoras' theorem with the recent work on sine, cosine and tangent. For example, give pupils a right-angled triangle with the opposite and hypotenuse stated and ask them to find the adjacent side. The obvious method would be to use Pythagoras' theorem. Show that the trigonometric approach gives the same answer, namely to find the angle θ using inverse sine, then use the cosine of the angle and the hypotenuse to find the adjacent side.

Incorporating exercise:	12L, 12M, 12N, 12P
Homework:	12.10a, 12.10b, 12.10c, 12.10d
Examples:	12.10a, 12.10b, 12.10c, 12.10d

Key words

angle of depression	isosceles triangle
angle of elevation	three-figure bearing
bearing	trigonometry

Learning objective(s)

- solve practical problems using trigonometry
- solve problems using an angle of elevation or an angle of depression
- solve bearing problems using trigonometry
- use trigonometry to solve problems involving isosceles triangles

Prior knowledge

Pupils should have covered Sections 12.4–12.9 and be familiar with calculating missing sides and angles of right-angled triangles using the trigonometric ratios sine, cosine and tangent. They should be able to recognise which trigonometric ratio is required in which situation.

Starters

- Pupils should already know which trigonometric ratio is required according to the question set, so test their speed and ability to do this. Draw a right-angled triangle on the board, giving an angle, stating a side, and labelling an unknown side as x. Ask them which ratio they would use: sin, cos or tan? What would be the equation they should write down? Change the angle and/or sides and repeat the question several times.
- Demonstrate examples of angles of elevation and depression using diagrams. If the angle of elevation from A to B is 65°, what is the angle of depression from B to A, and why?
- Demonstrate examples of bearings using diagrams, drawing attention to bearings in different quadrants. A ship sails from A to B on a bearing of 65°. Ask pupils to draw a diagram of the situation. What is the bearing from B to A, and why?
- Remind pupils of the properties of isosceles triangles. Give them an example of an isosceles triangle such as a triangle with sides 8 cm, 8 cm and 5 cm. How can you use Pythagoras' theorem to calculate its area?

Main teaching points

The first part of this section comprises some basic practical situations where right-angled triangles might occur. These generally involve diagonals of rectangles, or ladders leaning against walls. The difference with previous sections is that pupils will be expected to extract their own right-angled triangle from the information given in the question and correctly label its sides, before embarking on the main calculation.

This section includes three further topics covering:
- angles of elevation and depression
- trigonometry and bearings
- further problems involving isosceles triangles.

Angles of elevation and depression are merely descriptors for angles measured above and below a horizontal line. Give pupils one or two basic examples in the form of diagrams, emphasising the fact that the angles are measured from the horizontal.

Bearings, on the other hand, are measured from the vertical. Emphasise the three key points regarding bearings; namely that each bearing is an angle measured from a north line, in a clockwise sense, and should be written down using three digits. In most questions of this type, pupils will likely have to draw in their own north line at various places according to what the question might be asking.

The final subsection on isosceles triangles complements earlier work in Section 12.3 where Pythagoras' theorem was used when bisecting an isosceles triangle, forming a right-angled triangle, and finding a missing side. Almost the same procedure is followed here: an isosceles triangle is bisected to form a right-angled triangle upon which trigonometry can then be used to find a missing side or angle.

Differentiation

Throughout this section, being able to extract the information required in the form of right-angled triangles is crucial. It is this aspect that will undoubtedly separate the pupils in terms of ability, and therefore it is important to demonstrate as many examples as needed, particularly so in the case of bearings.

Plenary

These are a couple of more difficult questions involving trigonometry, with an emphasis on extracting right-angled triangles.

⦿ The diagram shows the cross-section of a cylindrical storage tank. Find the depth of the liquid above the bottom of the tank. (Answer: 1.28 m)

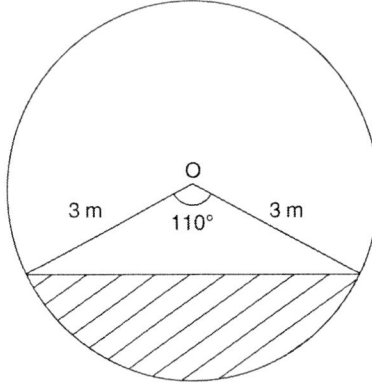

⦿ In order for the height of a tower to be measured, the angles of elevation of the top of the tower are taken from two points A and B, a distance of 130 m apart. The angle of elevation of the tower from A is 15° and the angle of elevation from B is 30°. Find the height of the tower. (Answer: 65 m)

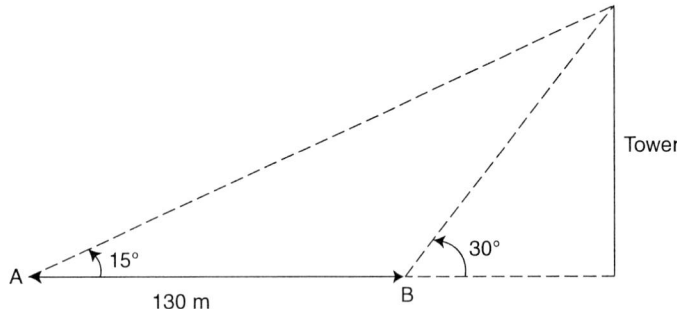

         Module 5: Algebra and Space, shape and measure

This chapter covers the geometry required at higher level, focusing on the interior and exterior angles of polygons and the associated formulae, along with a comprehensive look at the circle theorems.

Context

Although the circle theorems have few 'practical' applications in the usual sense of the word, they do provide pupils with important opportunities to focus on mathematical reasoning and proof. Proofs of all the circle theorems should be accessible to higher level candidates. Pupils are not required to reproduce these as part of the GCSE examination, however they should be prepared to offer logical reasons for their choices of values of missing angles, and the lower achieving pupils should be dissuaded from 'guessing' missing angles.

AQA B references

AO3 Shape, space and measures: Geometrical reasoning

13.1 3.2d "recall the definitions of special types of quadrilateral, including square, rectangle, parallelogram, trapezium and rhombus; classify quadrilaterals by their geometric properties"

13.2 3.2e "calculate and use the sums of the interior and exterior angles of quadrilaterals, pentagons and hexagons; calculate and use the angles of regular polygons"

13.3–13.6 3.2i "recall the definition of a circle and the meaning of related terms, including centre, radius, chord, diameter, circumference, tangent, arc, sector and segment; understand that the tangent at any point on a circle is perpendicular to the radius at that point; understand and use the fact that tangents from an external point are equal in length; explain why the perpendicular from the centre to a chord bisects the chord; understand that inscribed regular polygons can be constructed by equal division of a circle; prove and use the facts that the angle subtended by an arc at the centre of a circle is twice the angle subtended at any point on the circumference, the angle subtended at the circumference by a semicircle is a right angle, that angles in the same segment are equal, and that opposite angles of a cyclic quadrilateral sum to 180 degrees; prove and use the alternate segment theorem"

Route mapping

Exercise	D	C	B	A	A*
A	1–3	4–7			
B		all			
C			1–4	5–6	7
D			1	2–6	7
E			1–4	5	6
F				1–4	5

Answers to diagnostic Check-in test

1 35°	**2** 115°	**3** 36°
4 39°	**5** 82°	**6** 110°
7 50°	**8** 70°	**9** 69°

Find the value of x in each of these diagrams.

1

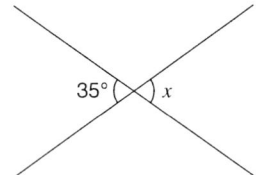

2

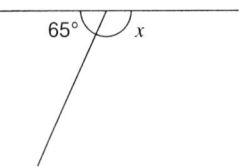

3

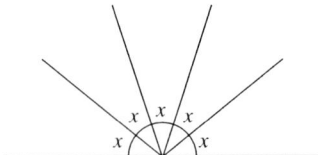

4

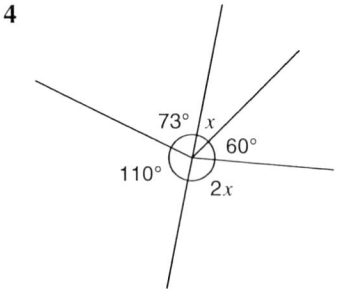

5

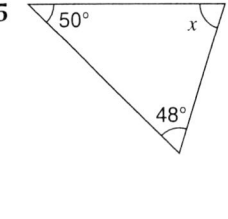

6

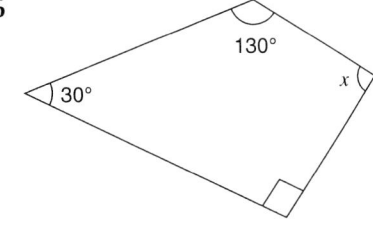

7

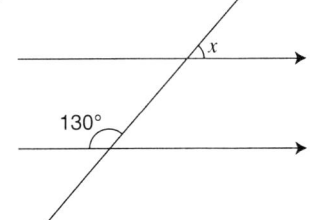

8

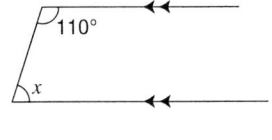

9

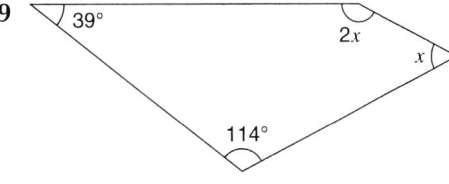

            Module 5: Algebra and Space, shape and measure

			Key words	
Incorporating exercise:	13A		equilateral	kite
Homework:	13.1		triangle	parallelogram
Example:	13.1		isosceles	rhombus
			triangle	trapezium

Learning objective(s)

● find angles in triangles and quadrilaterals

Prior knowledge

Pupils should know that the angles in a triangle add to 180°, and that in an isosceles triangle, two sides and two angles are the same. They should know that the angles in a quadrilateral add to 360°. They should also know that when a transversal cuts parallel lines the alternate angles are equal, the corresponding angles are equal and allied angles are supplementary.

Starter

Draw two parallel lines and a transversal. Ask pupils to label a pair of alternate angles, a pair of corresponding angles and a pair of allied angles. Ask them what properties these angles have.

Check pupils know that the angles in a triangle add up to 180°. Ask them what the angles in a quadrilateral add up to, and why. (A quadrilateral can be split into two triangles; so 2 × 180 = 360°.)

Main teaching points

Pupils should have already come across some, if not all, of the four quadrilaterals considered here: the parallelogram, trapezium, rhombus and kite. Yet, many pupils may be unable to describe them precisely. They could be asked to define what each shape is in turn. Try to sharpen their responses and elicit accurate definitions. For example:

● a trapezium is a quadrilateral with two parallel sides
● a parallelogram is a quadrilateral with two pairs of parallel sides
● a rhombus is a parallelogram with sides of equal length
● a kite is a quadrilateral with two pairs of equal adjacent sides.

Once these are established, the important angle properties of the four quadrilaterals can be taught.

● In a trapezium, the angles at the ends of each non-parallel side are supplementary (since they are allied angles).
● A parallelogram's opposite angles are equal.
● The diagonals of a rhombus bisect each other at 90°.
● A kite's longer diagonal bisects its shorter diagonal at right angles.

Plenary

Draw one each of an isosceles triangle, trapezium, parallelogram, kite and rhombus on the board (without marking identical angles and sides). Ask pupils (in turn) to mark pairs of identical angles or sides on the diagrams until all possibilities are exhausted.

Incorporating exercise:	13B
Homework:	13.2
Examples:	13.2

Key words

exterior angle	octagon
heptagon	pentagon
hexagon	polygon
interior angle	regular polygon

Learning objective(s)

● find interior angles and exterior angles in a polygon

Prior knowledge

Pupils should know that the interior angles of a triangle add to 180° and the interior angles in a quadrilateral add to 360°.

Starter

Ask pupils, "What is a 3-sided shape called; a 4-sided shape; a 5-sided shape;...a 12-sided shape?"

Main teaching points

Each polygon of n sides can be split into $n - 2$ triangles. The sum of the interior angles of a polygon of n sides is therefore given as $180(n - 2)$.

It is worth noting that this result is true for both convex and concave polygons. (A polygon is concave if and only if at least one of its internal angles is greater than 180°, so a concave polygon must have at least four sides.)

For a regular polygon of n sides, the size of each exterior angle is $\dfrac{360}{n}$.

The interior angle and exterior angle of a polygon add to 180°. The size of each interior angle is therefore $180 - \dfrac{360}{n}$.

The sum of all interior angles is therefore $n\left(180 - \dfrac{360}{n}\right) = 180n - 360 = 180(n - 2)$, verifying the opening teaching point.

Plenary

Ask pupils to complete this table for regular polygons:

Regular polygon	Number of sides	Exterior angle	Interior angle
Triangle	3		
Square	4		
Pentagon	5		
Hexagon	6		
Heptagon	7		
Octagon	8		
Nonagon	9		
Decagon	10		
Dodecagon	12		
n-gon	n		

What happens to the interior angle of a regular polygon as the number of sides increases? Why?

Incorporating exercise:	13C	Key words	
Homework:	13.3	arc	segment
Example:	13.3	circle	semicircle
		circumference	subtended
		diameter	

Learning objective(s)

● find angles in circles

Prior knowledge

Pupils should know that the interior angles of a triangle add to 180° and the interior angles in a quadrilateral add to 360°. They should be familiar with circle terms such as chord, segment, circumference, diameter, radius and arc. They should also be familiar with the angle properties of isosceles triangles.

Starter

Draw a suitably labelled circle in order to remind pupils of the main circle terms. For example, a tangent line could be drawn to the circle and the pupils asked to define precisely what a tangent line is.

Similarly ensure that pupils understand the terms chord, segment, circumference, diameter, radius and arc.

The word 'subtends' can also be usefully clarified with an example, before commencing the chapter proper.

Main teaching points

Circle theorem 1 states that the angle at the centre of a circle is twice the angle at the circumference subtended by the same arc. It is also worth pointing out to pupils the case where angle $\angle AOB$ is reflex, as pupils generally have more difficulty recognising this situation.

Circle theorem 2 simply states the angle in a semicircle is a right-angle. This is simply a special case of Circle theorem 1, where the angle at the circumference is 90°.

Circle theorem 3 states that angles at the circumference are all equal if they stand on the same arc. So, in this diagram, $q = r$.

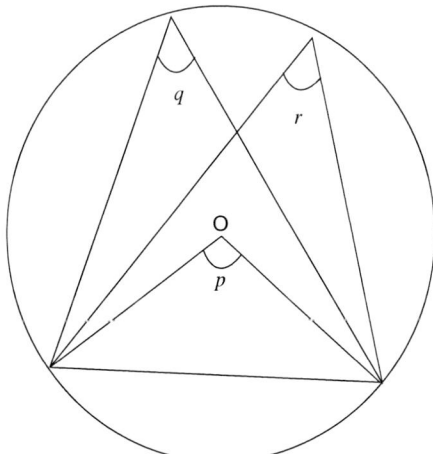

Formal proofs of each theorem are given in the plenary (see below). However, before launching into the proofs, it might be best to allow pupils to explore the use of the theorems by looking at particular examples. This will help them to recognise the situations where each might be applied. It is particularly important that pupils get to practise the application of Circle theorem 1, where the angle at the centre of the circle is reflex.

Circle theorem 1 is the key to understanding this work. After the proof of Circle theorem 1 is considered, the arguments, if not the formal proofs of the other two theorems, should be more easily understood.

Plenary

The circle theorems 1, 2 and 3 may be proven from the properties of an isosceles triangle. It is suggested that these proofs are given to the pupils. You could either give all proofs as a demonstration or set proofs for theorems 2 and 3 as a plenary exercise after Circle theorem 1 has been established.

Circle theorem 1: The angle subtended by a chord at the centre of a circle is twice the angle subtended at the circumference.

This proof is essentially given as an exercise in question 7 of Exercise 7C in the Pupil Book, which may already have been completed by the more able pupils. The proof starts by considering this diagram:

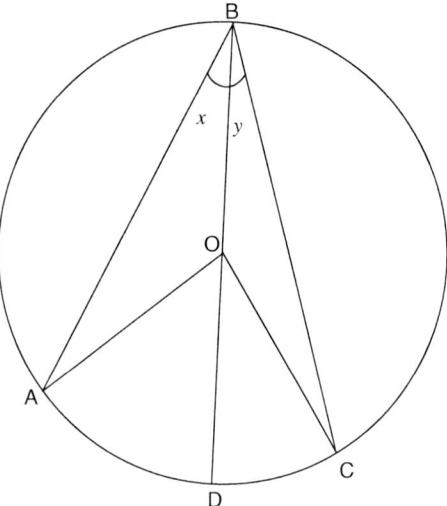

Since AOB and BOC are isosceles triangles, we have angle $\angle BAO = x$ and $\angle BCO = y$. Therefore $\angle BOA = 180 - 2x$ and $\angle BOC = 180 - 2y$.

Now since angles on a straight line add up to 180°, we have $\angle AOD = 2x$ and $\angle DOC = 2y$, therefore $\angle AOC = 2x + 2y = 2(x + y) = 2\angle ABC$ as required.

Circle theorem 2: Every angle at the circumference of a semicircle that is subtended by the diameter of the semicircle is a right angle.

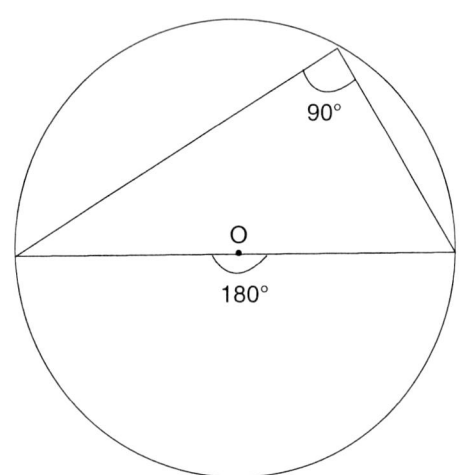

The proof follows immediately from Circle theorem 1. Let the chord be the diameter in this case. Then the angle at the chord is 180° and so the angle at the circumference must be 90°.

Circle theorem 3: Angles at the circumference in the same segment of a circle are equal.

Consider this diagram:

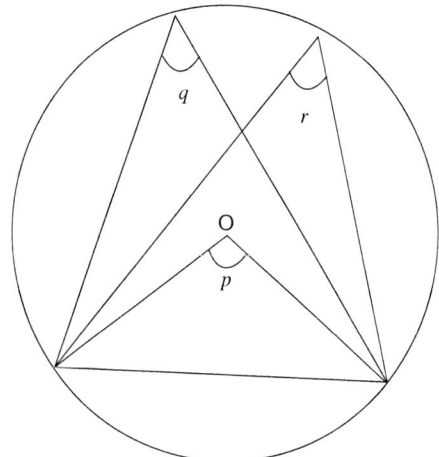

From Circle theorem 1, we know $p = 2q$ but also $p = 2r$.
So, $2q = 2r$ and $q = r$ as required.

     Module 5: Algebra and Space, shape and measure

Incorporating exercise: 13D	**Key words**
Homework: 13.4	cyclic
Example: 13.4	quadrilateral

Learning objective(s)

● find angles in cyclic quadrilaterals

Prior knowledge

Pupils should have covered Section 13.3 on finding angles in circles.

Starter

It might prove beneficial in following straight on from the proofs offered in the previous section's plenary to illustrate the proof of Circle theorem 4 to pupils (given in the main teaching points below). However, this could always be left until later as a follow-up to question 7 in Exercise 13D of the Pupil Book.

Main teaching points

A cyclic quadrilateral is defined as being a quadrilateral whose vertices lie on the circumference. Circle theorem 4 states that the opposite sides in a cyclic quadrilateral are supplementary: that is, $a + c = 180°$ and $b + d = 180°$.

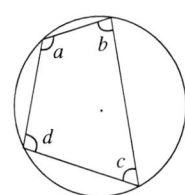

The proof of Circle theorem 4 follows from this diagram:

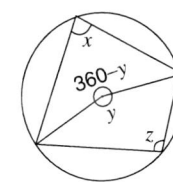

From Circle theorem 2, we know $y = 2x$. Therefore the reflex angle at the centre of the circle is $360 - y = 360 - 2x$.

Using Circle theorem 2 again with regard to this reflex angle, we see that $360 - 2x = 2z$. So, $z = 180 - x$ and $x + z = 180°$ as required.

Plenary

A further interesting result with regard to cyclic quadrilaterals is that of Brahmagupta's formula for the area of a quadrilateral. When the quadrilateral is cyclic, Brahmagupta's formula reduces to a direct analogy of Hero's formula for triangles.

Given a cyclic quadrilateral with sides a, b, c, d, the area is given by:

$$A = \sqrt{s(s-a)(s-b)(s-c)(s-d)} \quad \text{where} \quad s = \frac{a+b+c+d}{2}$$

Although the proof is very lengthy and not particularly inspiring, pupils might like to verify Brahmagupta's formula in the case of an inscribed square.

			Key words
Incorporating exercise:	13E		chord
Homework:	13.5		point of contact
Example:	13.5		radius
			tangent

Learning objective(s)

● find angles in circles when tangents or chords are used

Prior knowledge

Pupils should have covered Sections 13.3–13.4 on angles in circles and cyclic quadrilaterals.

Starter

The first main teaching point is Circle theorem 5, that a tangent line is perpendicular to the radius at the point of contact with the circle. It is suggested the proof of this is gone through with pupils as a whole.

Main teaching points

Circle theorem 5 states that a tangent to a circle is perpendicular to the radius drawn to the point of contact. The proof requires little mathematical knowledge other than the fact that the hypotenuse is the longest side in a right-angled triangle. It is also a straightforward example of proof by contradiction.

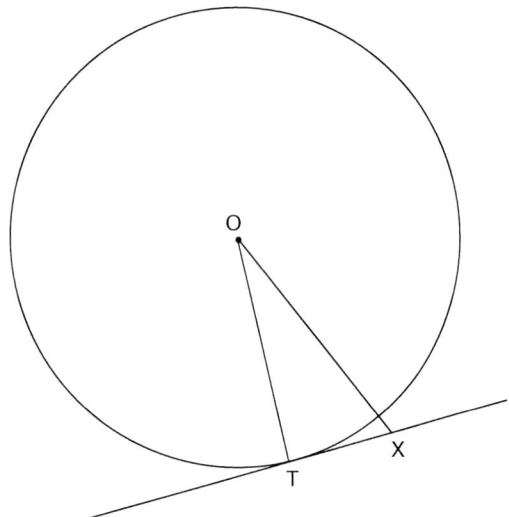

The diagram above shows a circle centre O and a tangent line. Suppose OT is not perpendicular. Then we can draw a line that IS perpendicular. Suppose this is OX, so that $\angle OXT = 90°$. Then OXT is a right-angled triangle, with OT being the hypotenuse. However if OT is the longest side, this must force X to lie within the circle. But this would then mean the tangent line would cross the circle in two places, which is clearly a contradiction. Therefore OT must be perpendicular as required.

Although accessible for the more able pupils, the proofs for the other three circle theorems covered in this section (theorems 6–8) rely on the idea of congruent triangles, which is not covered until Section 14.1 of the Pupil Book. You could decide to return to the proofs later, or choose to cover 14.1 in conjunction with this section.

The proofs for circle theorems 6 and 7 follow from the diagram on the next page. In the diagram, AB and BC are two tangent lines and triangles OAB and OCB are congruent.

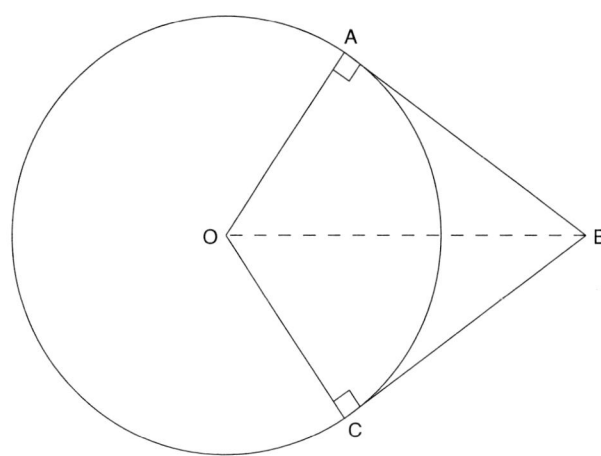

From the tangent property, we know ∠OAB = ∠OCB = 90°. Side OB is common to both triangles, and so by the RHS rule, triangles OAB and OCB are congruent.

Therefore, AB = CB also follows due to their being corresponding sides (this is Circle theorem 6). Similarly, it follows from the congruence property that ∠ABO = ∠CBO (this is Circle theorem 7).

To prove, Circle theorem 8, that a perpendicular line drawn to a chord from the centre of a circle bisects the chord, consider this diagram.

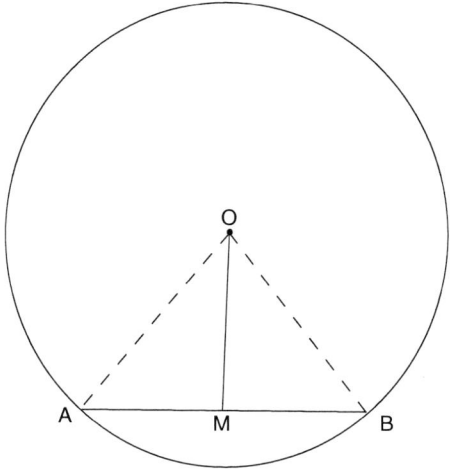

We are given that ∠AMO = ∠BMO = 90°, and OA = OB and OM is common to both triangles. Therefore, by the RHS rule, triangles OAM and OBM are congruent, and so AM = BM as they are corresponding sides. So OM bisects AB.

Plenary

Discuss with pupils the converse of Circle theorem 8: that a line drawn from the centre of a circle to the midpoint of a chord is perpendicular to the chord. Is this always true? Can it be proven?
The proof again makes use of the idea of congruent triangles. It follows from this diagram, where M is the midpoint of AB.

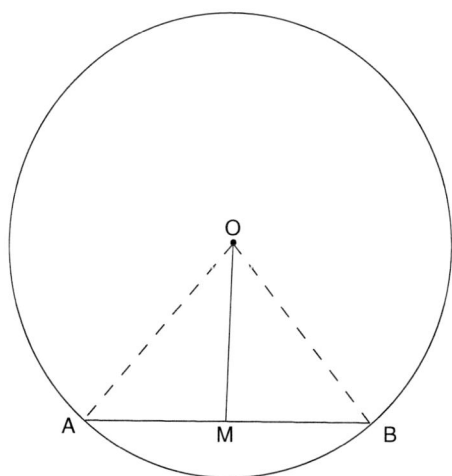

Consider triangles OAM and OBM. AM = MB since M is the midpoint. OA = OB since they are both radii and OM is common to both triangles. Therefore OAM and OBM are congruent by the SSS rule.

The corresponding angles, therefore, are also the same, so ∠OMB = ∠OMA. Furthermore, since they lie on a straight line, their sum must be 180° and so they must each be 90° as required.

Incorporating exercise:	13F		Key words
Homework:	13.6		alternate
Example:	13.6		segment
			chord
			tangent

Learning objective(s)

• find angles in circles using the alternate segment theorem

Prior knowledge

Pupils should have covered circle theorems 1–8 from Sections 13.3–13.5.

Starter

It is suggested that you take pupils through the proof of Circle theorem 9. The proof usefully includes circle theorems 3 and 5, so acts as a reminder of previous work.

Main teaching points

Circle theorem 9 states that the angle between a tangent and a chord through the point of contact is equal to the angle in the alternate segment: that, in other words, $x = y$, in this diagram.

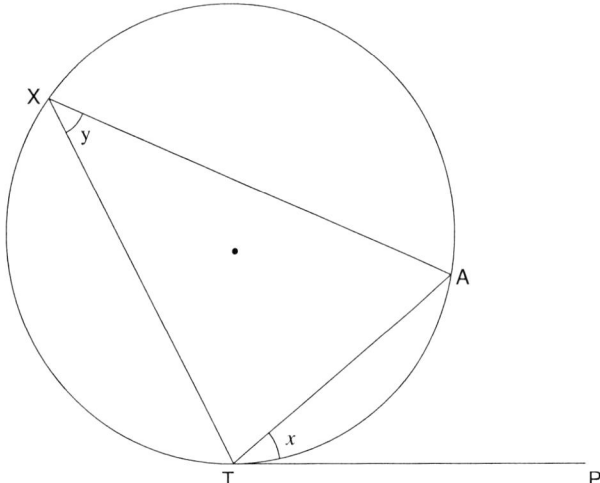

To prove the theorem, consider the situation where TX is perpendicular to the tangent line TP shown in this next diagram.

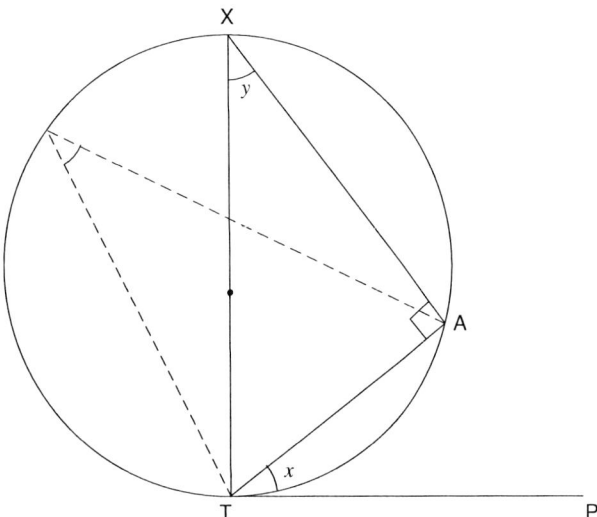

As XT is a diameter, $\angle XAT = 90°$. Since XT and the tangent line TP are perpendicular, we have $\angle XAT = 90 - x$, therefore $y = x$.

Now since angles in the same segment are equal, it does not matter where X is placed on the arc of the alternate segment. The result will still hold.

Plenary

Circle theorem 9 is true for obtuse angles. Pupils might like to try proving it for themselves by using this diagram.

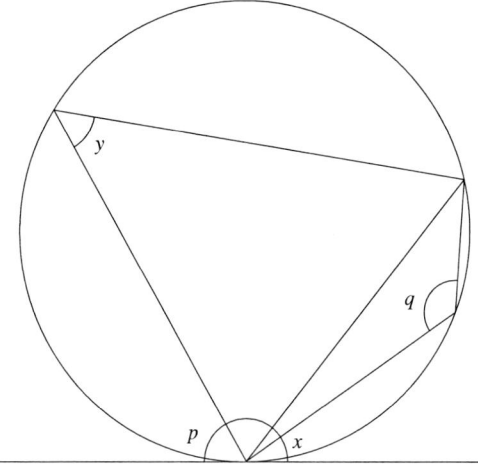

Circle theorem 9 states that $x = y$. We also know that $x + p = 180°$ since both angles lie on a straight line. In addition, $q + y = 180°$ since they are opposite angles in a cyclic quadrilateral.

So $p = 180 - x = 180 - y = 180 - (180 - q) = q$ as required.
Therefore, Circle theorem 9 holds for both acute and obtuse angles.

Overview

14.1 Congruent triangles
14.2 Translations
14.3 Reflections
14.4 Rotations
14.5 Enlargements
14.6 Combined transformations

This chapter covers two main sections in transformation geometry. The first is the work on congruent triangles with the corresponding rules for congruency. The remainder of the chapter devotes itself to the work on transformations required at this level, namely translations, reflections, rotations and enlargements.

Context

This chapter is geometrical in nature and so pupils are likely to recognise few 'practical' applications other than perhaps reflection in relation to using a mirror. In fact, transformations are important and occur everywhere around us – from facial symmetry in biology through to the idea of isometries in chemistry. Islamic art is particularly notable for its highly geometric designs and dependence on mathematics and geometric transformations.

AQA B references

AO3 Shape, space and measures: Geometrical reasoning

14.1   3.2f "understand and use SSS, SAS, ASA and RHS conditions to prove the congruence of triangles using formal arguments..."

AO3 Shape, space and measures: Transformations and coordinates

14.2–14.6   3.3a "understand that rotations are specified by a centre and an (anticlockwise) angle; use any point as the centre of rotation; measure the angle of rotation, using right angles, fractions of a turn or degrees; understand that reflections are specified by a (mirror) line; understand that translations are specified by giving a distance and direction (or a vector), and enlargements by a centre and a positive scale factor"

14.2–14.6   3.3b "recognise and visualise rotations, reflections and translations including reflection symmetry of 2-D and 3-D shapes, and rotation symmetry of 2-D shapes; transform triangles and other 2-D shapes by translation, rotation and reflection and combinations of these transformations; use congruence to show that translations, rotations and reflections preserve length and angle, so that any figure is congruent to its image under any of these transformations; distinguish properties that are preserved under particular transformations"

14.5   3.3c "recognise, visualise and construct enlargements of objects; understand from this that any two circles and any two squares are mathematically similar, while, in general, two rectangles are not, then use positive fractional and negative scale factors"

14.5   3.3d "recognise that enlargements preserve angle but not length; identify the scale factor of an enlargement as the ratio of the lengths of any two corresponding line segments; understand the implications of enlargement for perimeter;..."

Route mapping

Exercise	D	C	B	A	A*
A			1–4	5–6	
B		all			
C	1–4	5–9			
D	1	2–9			
E		1–4	5		
F		1–2	3–9		

Answers to diagnostic Check-in test

1 a 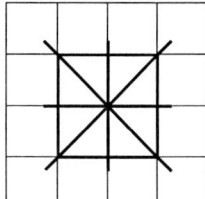 b 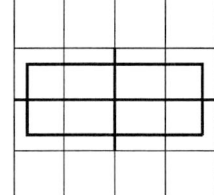 c 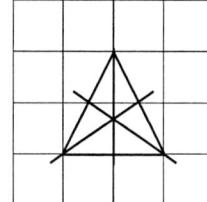 d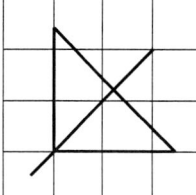

2 a 4   b 2   c 3   d 1

3 a (3, –1)   b (3, 3)   c (3, –3)

4 a $y = 3$ b $x = –2$ c $y = x$ d $y = –x$

1 Draw the lines of symmetry on the following shapes.

1 a 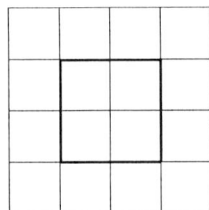   **b** 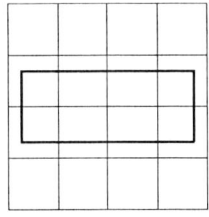   **c** 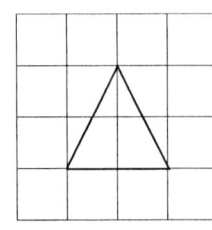   **d**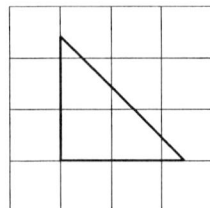

2 Give the order of rotational symmetry of the shapes in question 1.

3 a State the coordinates of the point of intersection of the lines
$x = 3$ and $y = -1$.

b State the coordinates of the point of intersection of the lines
$x = 3$ and $y = x$.

c State the coordinates of the point of intersection of the lines
$x = 3$ and $y = -x$.

4 Write down the equations of the lines drawn on the grid.

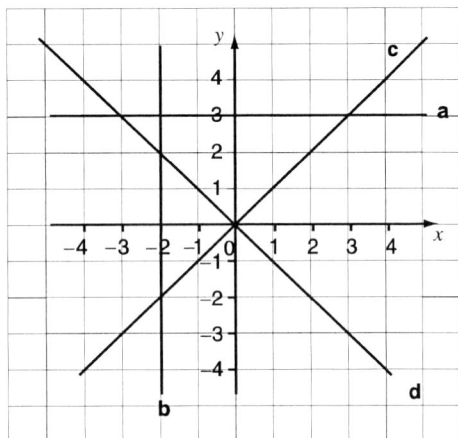

Equation of line **a** is _____

Equation of line **b** is _____

Equation of line **c** is _____

Equation of line **d** is _____

           Module 5: Algebra and Space, shape and measure

Incorporating exercise:	14A	Key words
Homework:	14.1	congruent
Examples:	14.1	

Learning objective(s)

● show that two triangles are congruent

Prior knowledge

Pupils should be able to recognise congruent shapes and understand the meaning of congruency.

Starter

Remind pupils of the definition of congruent shapes; that is, identical shapes. Draw these two triangles on the board and ask pupils if the shapes are congruent. Most will agree.

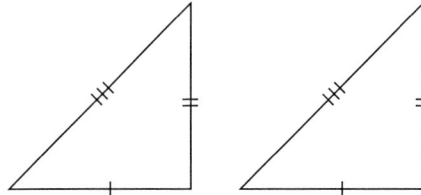

Then draw another set of triangles on the board, one a reflection of the other. There may well be some indecision and argument between pupils about whether this pair are congruent.

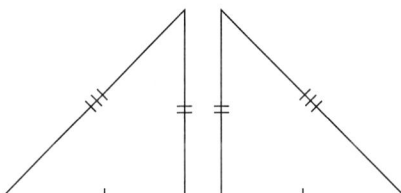

This simple exercise can then be used to emphasise that we need to be clear about what is meant by congruent or identical. Emphasise that it does not matter that the triangle is reflected – the sides and angles remain identical and we therefore class these triangles as congruent.

Main teaching points

Congruent triangles are identical triangles: they have exactly the same shape and size. One of four conditions is sufficient for two triangles to be congruent.

Condition 1: The three sides of one triangle are equal to the three sides of the other triangle (the SSS rule).

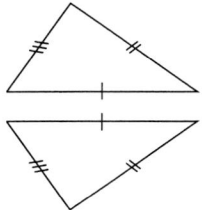

Condition 2: Two sides and the included angle of one triangle are equal to two sides and the included angle of the other (the SAS rule).

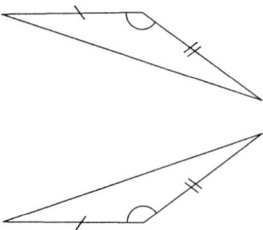

Condition 3: Two angles and a side of one triangle are equal to two angles and the corresponding side in the other (the ASA rule).

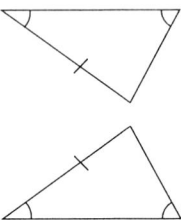

Condition 4: For right-angled triangles, the hypotenuse and another side are equal to the hypotenuse and a second side in the other triangle (the RHS rule).

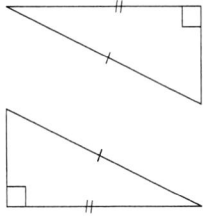

Common mistakes

Pupils often think that it is sufficient to show that the three sets of angles are the same in both triangles to prove congruency. This is untrue, of course, and merely shows that the two triangles are similar. It is important therefore to reinforce continually the four conditions that establish congruency.

Plenary

Ask pupils to consider whether these triangles are congruent.

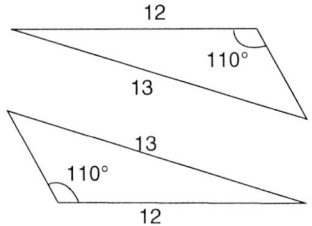

The answer is not obvious, as seemingly none of the congruency rules initially fit. However, the remaining angles to be found must both be acute, and so there will be no ambiguity when calculating the non-included angle, and hence none with regards to the included angle. These triangles are therefore congruent due to the SAS rule.

Incorporating exercise:	14B
Homework:	14.2
Examples:	14.2

Key words
transformation   vector
translation

Learning objective(s)

- translate a 2-D shape

Prior knowledge

There is very little prior knowledge needed in order to begin basic work on transformations. Some recent experience of work involving Cartesian coordinates might be useful, but is not essential.

Starter

Begin by discussing the general idea of the transformation of an object, and what this means. Many pupils will have come across different types of transformation before, so ask them to name specific types of transformation.

Emphasise that the four basic transformations – translation, reflection, rotation and enlargement – will each be looked at in turn, and that pupils need to be able to recognise, describe and perform these transformations.

Main teaching points

In a translation, the whole object is moved in the same direction. Its size does not change, and therefore the image of the object after translation is congruent to the original object.

A translation is described using a column vector in the form $\begin{pmatrix} x \\ y \end{pmatrix}$ where x indicates how far to the right the object

must translate, and y indicates how far up. Therefore, if the object is to move to the left, x will be negative.

Similarly, if the object moves downwards, y will be negative.

If a translation $A \rightarrow B$ is described by the column vector $\begin{pmatrix} x \\ y \end{pmatrix}$, then $B \rightarrow A$ is described by $\begin{pmatrix} -x \\ -y \end{pmatrix}$.

Common mistakes

The terms transformation and translation can often be confused by lower achieving pupils. Ensure that pupils understand that a translation is a specific type of transformation, as is a rotation and a reflection.

Plenary

An object A is translated through $\begin{pmatrix} a \\ b \end{pmatrix}$. Call its image B. Image B is then translated through $\begin{pmatrix} c \\ d \end{pmatrix}$ and its image

called C. What column vector describes the translation $A \rightarrow C$? What column vector describes the translation $C \rightarrow A$?

Incorporating exercise:	14C	Key words	
Homework:	14.3	image	object
Examples:	14.3	mirror line	reflection

Learning objective(s)

- reflect a 2-D shape in a mirror line

Prior knowledge

Pupils should have completed Section 14.2 on translation. They should also be familiar with equations of lines of the type $y = x$, $y = -x$, $y = a$ and $x = b$.

Starter

Draw a set of axes from -5 to $+5$ for x- and y-axes.

Draw some horizontal and vertical lines and ask pupils for their equations.

Make sure they understand that x lines are vertical and y lines are horizontal.

Finish with $y = x$ and $y = -x$.

Main teaching points

A reflection transforms a shape so that it becomes a mirror image of itself. After a reflection, the object's size does not change, and therefore the image of the object after translation is congruent to the original object.

To reflect a 2-D shape, it is sometimes easier to reflect each vertex of the shape in turn, before joining the points obtained to form the image object. Each vertex is reflected perpendicular to the mirror line and is the same distance from it.

To describe a reflection, all that is needed is to describe the mirror line in which the object is reflected. This could be a simple mirror line drawn and labelled in a given diagram or, and more likely, it could be a standard line equation of the form $y = x$, $y = -x$, $y = a$ or $x = b$ if the reflection takes place using Cartesian axes.

Plenary

After a reflection in a given line, a set of coordinates is transformed as follows $(3, 1) \rightarrow (2, 0)$, $(3, 3) \rightarrow (0, 0)$, $(1, 5) \rightarrow (-2, 2)$, $(5, 1) \rightarrow (2, -2)$, $(2, 3) \rightarrow (0, 1)$.

Draw the mirror line. What is its equation? (Answer: $x + y = 3$)

What would be the coordinates of the image of (a, b) under a reflection in this line? (Answer: $(3 - b, 3 - a)$)

Incorporating exercise:	14D	
Homework:	14.4	
Examples:	14.4	

Key words

angle of rotation   clockwise
anticlockwise   rotation
centre of
     rotation

Learning objective(s)

● rotate a 2-D shape about a point

Prior knowledge

Pupils should ideally have covered Sections 14.2 and 14.3 on translations and reflections.

Starter

Ask pupils to draw a T-shape.

They should then draw another T-shape to add to the first, to make a shape with rotational symmetry of order 2.
Ask: "Who has a shape like this?

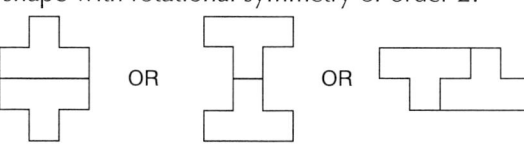

Ask pupils what the difference is between the diagrams.

Establish the concept of the centre of rotation. Mark the centre on the diagrams.

Main teaching points

A rotation turns a shape about a fixed point called the centre of rotation. After a rotation, the object's size does not change, and therefore the image of the object after rotation is congruent to the original object.

To describe a rotation fully, you need to give the centre of rotation (usually as a coordinate), the sense of the rotation (clockwise or anticlockwise) and the angle of rotation (most likely 90°, 180° or 270°).

In order to rotate a shape through a given angle, say 90°, it is often best to use tracing paper. Trace the object then, keeping the pencil point fixed at the centre of rotation, rotate the tracing paper through 90°. Retrace the object to complete the rotation. With practise, it should become easier to 'see' where the image should finish up, especially when using Cartesian axes.

Plenary

The point (2, 4) is rotated 90° anticlockwise about the centre of rotation (0, 0). Can pupils give the coordinates of the image point without resorting to drawing axes? (Answer: (–4, 2))

The point (2, 4) is rotated 180° about the centre of rotation (0, 0). Can pupils give the image point?
(Answer: (–2, –4))

The point (2, 4) is rotated 270° anticlockwise about the centre of rotation (0, 0). Can pupils give the image point?
(Answer: (4, –2))

Repeat for a general point (*a*, *b*). What will its coordinates be after rotating **a** 90° anticlockwise, **b** 180° and **c** 270° anticlockwise?

Incorporating exercise:	14E
Homework:	14.5
Examples:	14.5

Key words

centre of	enlargement
enlargement	scale factor

Learning objective(s)

○ enlarge a 2-D shape by a scale factor

Prior knowledge

Pupils should ideally have covered Sections 14.2–14.4 on translations, reflections and rotations.

Starter

Start by asking pupils what is meant by enlarging a shape by a scale factor of 2. Most are likely to have the (correct) idea that the sides of the shapes will be doubled in length.

Then ask them what might be meant by enlarging a shape by a scale factor of $\frac{1}{2}$ or perhaps –2. At this stage there

may well be some uncertain answers. This might then provide a useful lead-in to an initial summary of the types of enlargement.

It is suggested that pupils be given examples of enlargements with a positive scale factor, a negative scale factor and a fractional scale factor.

Enlargement scale factor k, $k > 1$. Note that OA′ = 2OA, OB′ = 2OB and OC′ = 2OC, so $k = 2$. Note also that A′B′ = 2AB, B′C′ = 2BC and C′A′ = 2CA.

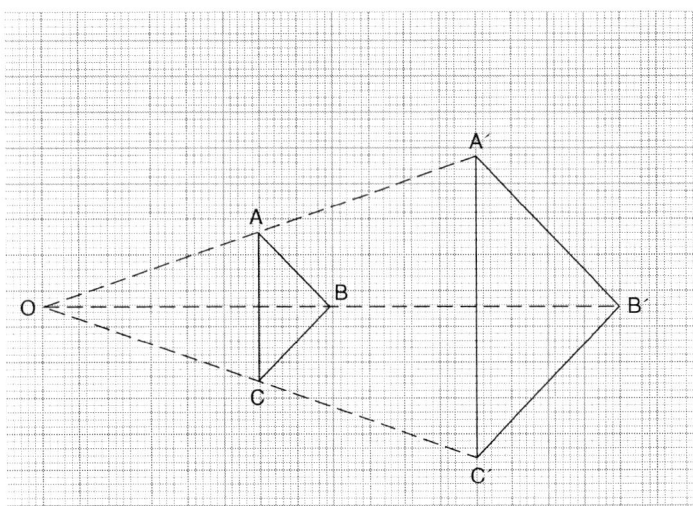

Module 5: Algebra and Space, shape and measure

Enlargement scale factor k, $0 < k < 1$. Note that $OA' = \frac{1}{4}OA$, $OB' = \frac{1}{4}OB$ and $OC' = \frac{1}{4}OC$, so $k = \frac{1}{4}$.

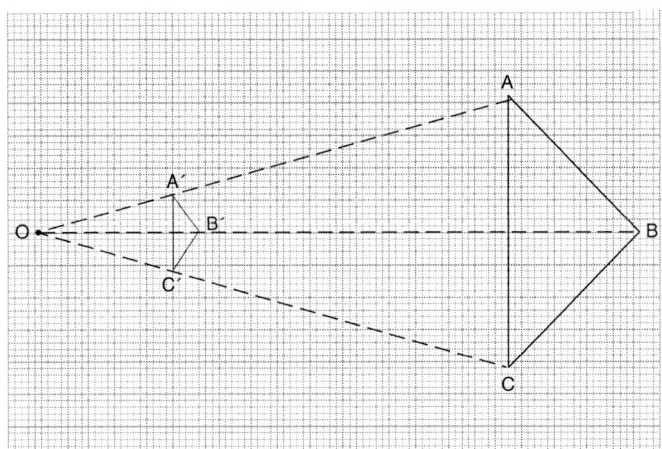

Enlargement scale factor k, $k < 0$. Note that $OA' = \frac{1}{2}OA$, $OB' = \frac{1}{2}OB$ and $OC' = \frac{1}{2}OC$. The image and object

are on opposite sides of the centre of enlargement. Therefore the image is inverted and $k = -\frac{1}{2}$.

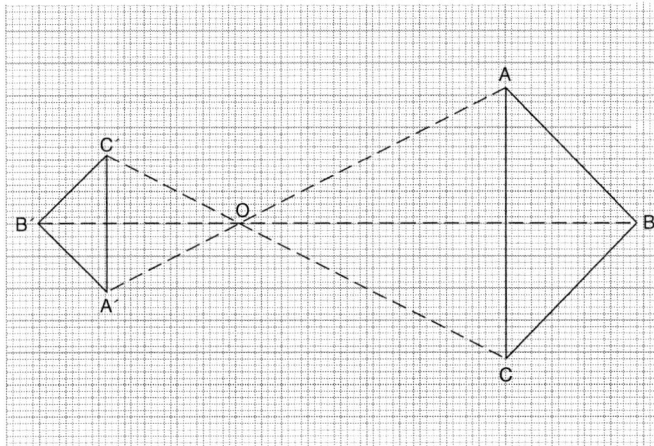

Main teaching points

An enlargement is a transformation that changes the size of an object. To describe an enlargement, you need a centre of enlargement and a scale factor.

Following an enlargement (and unless the scale factor is 1 or –1), the sides of the image of the original object will have increased or decreased in proportion. The object and image will not therefore be congruent, though they *will* be similar.

A negative scale factor k will cause an inverted image to be obtained. A scale factor k where $|k| < 1$ will cause the image to be reduced in size. Each of these transformations, however, are still described as enlargements.

To perform an enlargement of scale factor k, consider each vertex of the original object in turn. The image of each vertex should be k times as far from the centre of enlargement as it was originally. If k is negative, the new image point will be on the opposite side of the centre of enlargement. The centre of enlargement can be found by drawing straight lines through corresponding object and image vertices. They will all meet at the centre of enlargement. If any of them do not, then this is one way of showing the transformation cannot be an enlargement.

Plenary

Summarise in a table the information needed to describe each transformation studied so far.

Transformation	Information needed
Translation	column vector
Reflection	mirror line
Rotation	angle, sense (direction), centre of rotation
Enlargement	scale factor, centre of enlargement

Incorporating exercise:	14F	Key words	
Homework:	14.6	enlargement	transformation
Examples:	14.6	reflection	translation
		rotation	

Learning objective(s)

● combine transformations

Prior knowledge

Pupils should have covered Sections 14.2–14.5 on translations, reflections, rotations and enlargements.

Starter

Draw an x-axis and y-axis with two squares: square A with vertices at coordinates $(-2, 0)$, $(-2, 2)$, $(0, 0)$ and $(0, 2)$, and square B with vertices $(0, 0)$, $(0, 2)$, $(2, 2)$, $(2, 0)$. If B is a transformation of A, can pupils describe the transformation? Is it possible to describe it as a translation? Similarly, can it be described as a reflection, a rotation or an enlargement?

Main teaching points

This section is concerned with describing transformations that could be either translations, reflections, rotations or enlargements. Enlargements should be the first to be considered, as they are the only transformation that changes the size of the object and, hence, they are easily spotted. If an enlargement is then ruled out, the translation is the next easiest to spot.

When one transformation is followed by another, the result is sometimes equivalent to a single transformation. For example, the combination of two translations is equivalent to a single translation that can be obtained simply by adding the two column vectors. An enlargement about a given point followed by a translation is equivalent to an enlargement with the same scale factor, but using a different centre of enlargement. None of these specific examples need be learnt for GCSE; questions of this type will generally ask for transformations $A \rightarrow B$ and $B \rightarrow C$ to be carried out, before asking that the transformation $A \rightarrow C$ be identified.

Plenary

Question 2b in Homework sheet 14.6 asks for the single transformation equivalent to a reflection in the line $x = 3$ followed by a reflection in the line $x = 6$. The answer is a translation through $\begin{pmatrix} 6 \\ 0 \end{pmatrix}$.

What single transformation would be equivalent to a reflection in the line $x = a$ followed by a reflection in the line $x = b$ $(b > a)$? (Answer: translation through $\begin{pmatrix} 2(b-a) \\ 0 \end{pmatrix}$)

What single transformation would be equivalent to a reflection in the line $x = b$ followed by a reflection in the line $x = a$ $(b > a)$? (Answer: translation through $\begin{pmatrix} 2(a-b) \\ 0 \end{pmatrix}$)

Overview

This chapter is broadly split into two strongly linked topics. Sections 1 and 2 develop skills of construction (grades C and B). These help with the understanding of Sections 3 and 4, which cover drawing, describing and solving problems with loci.

Context

The material has applications in the accurate drawing of structures involving triangles, such as bridges and roofs. It also applies in any situation that requires determining the location of a facility, such as television, radio and telephone masts, radar stations, water sprinkler systems on golf courses and movement sensors on security lighting.

AQA B references

AO3 Shape, space and measures: Measures and construction

15.1–15.2 3.4c "use straight edge and compasses to do standard constructions including an equilateral triangle with a given side, the midpoint and perpendicular bisector of a line segment, the perpendicular from a point to a line, the perpendicular from a point on a line, and the bisector of an angle"

15.3–15.4 3.4e "find loci ... by reasoning ... to produce shapes and paths"

Route mapping

Exercise	D	C	B	A	A*
A		all			
B			all		
C		all			
D		all			

Answers to diagnostic Check-in test

1 **a** An equilateral triangle of side length 5 cm.
   b An equilateral triangle of side length 6 cm.
2 **a** 100 cm (1 m), 5 cm   **b** 150 m   **c** 6 km   **d** 200 cm (2 m)
   e 2000 cm (20 m)   **f** 50 000 cm (500 m or 0.5 km)   **g** 200 km (20 000 000 cm)

1 Using a pencil, ruler, protractor and a pair of compasses, draw:
  a a triangle with two sides of 5 cm and included angle of 60°.

  b a triangle with all three sides of length 6 cm.

2 This is a scale drawing of a rectangle.

10 cm

0.5 cm

  a If the scale is 1 cm : 10 cm, the rectangle is _____ long and _____ thick.

  b If the scale is 1 cm : 15 m, the rectangle is _____ long.

  c If the scale is 1 cm : 12 km, the rectangle is _____ thick.

  d If the scale is 1 : 20, the rectangle is _____ long.

  e If the scale is 1 : 4000, the rectangle is _____ thick.

  f If the scale is 1 : 100 000, the rectangle is _____ thick.

  g If the scale is 1 : 2 000 000, the rectangle is _____ long.

			Key words	
Incorporating exercise:	15A		angle bisector	perpendicular
Homework:	15.1		bisector	bisector
Example:	15.1		line bisector	

Learning objective(s)

● bisect a line and an angle

Prior knowledge

Pupils should be able to use compasses to draw arcs and circles accurately.

Starter

Start perhaps by showing the wrong way to bisect a line. Draw a straight line, measure half way along and make a mark. Use a protractor to make a mark at 90° above and below the line. Draw a line between the two marks.

Tell pupils that if they used this method to cut a line into two equal parts (bisect) at right angles (so a perpendicular bisector), they would lose *all* marks for this question.

Demonstrate on the board the correct method to bisect a straight line, using a pair of compasses and a straight edge.

Main teaching points

Pupils need to understand that bisect means to cut into two *equal* parts. When a line is bisected, we draw a perpendicular bisector: that is, the line that bisects is at 90° to the one that has been bisected.

As with any construction question, lack of accuracy means lack of marks.

Common mistakes

To rub out construction lines. Not using compasses.

Plenary

You will need a board pen attached to a piece of string (at least 50 cm long). Get a volunteer to bisect a 120° angle using the pen and the string. Then ask a second volunteer to bisect one of the resulting 60° angles, a third to bisect one of the 30° angles, and a fourth to check the size of the 15° angle with a board protractor.

Draw a horizontal line, length 40 or 50 cm, on the board. Give another volunteer the pen and the string, and ask him or her to bisect the line, and get a final volunteer to bisect one of the two halves.

Incorporating exercise:	15B
Homework:	15.2
Examples:	15.2

Key words
construct
perpendicular

Learning objective(s)

- construct perpendiculars from a point
- construct an angle of 60°

Prior knowledge

Pupils should be able to bisect a line.

Starter

Start by reviewing how to bisect a line. Draw a 7 cm line and, using compasses and a straight edge, bisect the line.

Main teaching points

Emphasise to pupils that constructing a perpendicular to and from a point uses the same essential method as the starter activity except for one additional action: that of making two equidistant marks on the line either side of the point by using a compass:

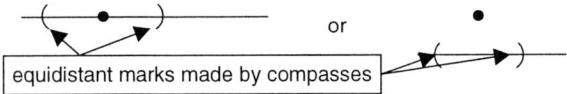

Common mistakes

Some pupils might bisect the line, or use a protractor to measure 90° from the point (rather than *construct* the perpendicular).

Plenary

Using a board pen attached to a piece of string (at least 50 cm long), get a volunteer to follow instructions from another pupil to construct a perpendicular from a point on a line.

Get two other pupils to construct a perpendicular from a point to a line; again, with one doing the drawing and one giving instructions.

Do the same again, by getting two other pupils to construct an equilateral triangle on the board of side length 20–30 cm.

Incorporating exercise:	15C
Homework:	15.3
Example:	15.3

Key words
loci
locus

Learning objective(s)

● draw a locus for a given rule

Prior knowledge

Pupils need to be able to construct the perpendicular bisector of a line (although it will usually be shown as two points).

Starter

Pose the following problems:

● Percy the goat is tethered in a field to a rope which is 10 m long. Show the region where Percy can go.

● Two radio masts situated 50 miles apart can both transmit up to 30 miles. Show the region where you can get transmissions from both masts.

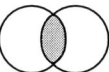

If appropriate, continue with these:

● Percy the goat (no longer in the field) has been tethered to a shed. The rope is 4 m long. Show where he can wander.

● Percy is now tied to a tree. He only likes to walk clockwise – show what would happen.

● Percy's owner has a bicycle with a reflective sticker on the rim of the front wheel. Show the path of the sticker as the wheel goes round.

● The back wheel has a sticker in the centre. Show its path as the wheel goes round.

● This line is the border of two countries – nobody is allowed within 1 km either side – how can we draw this?

Main teaching points

If a question asks for the distance from a single point, it usually means that a single circle from that point is required.

If the distance from two points is required, it usually means that pupils need to bisect the line joining the two points (by either drawing it in or imagining it is there).

Keeping the same distance from a line usually means drawing circles at the ends (and any bends) and joining the circles with straight lines.

Loci questions, such as the goat tethered to a shed or a tree, or the movement of a point on a bicycle wheel, require some imagination and practice.

Common mistakes

Pupils can confuse a distance from a point with the distance from a line. Lack of accuracy is a common mistake, and you should stress that ± 2 mm is the maximum error allowed in drawing and construction work.

Differentiation

Some pupils will probably benefit from having a piece of string to 'see' what happens from a fixed point.

Plenary

Ask pupils to explain (no hands allowed!) how to:

- show a point moving at a fixed distance from a fixed point
- show a point moving so that it is always the same distance (equidistant) from two fixed points
- show a point moving so that it is always at a fixed distance from a line.

 Module 5: Algebra and Space, shape and measure

Incorporating exercise:	15D
Homework:	15.4
Example:	15.4

Key words
loci
scale

Learning objective(s)

⚬ solve practical problems using loci

Prior knowledge

Pupils will need a sound knowledge of defining a locus, which is covered in Section 15.3.

Starter

Percy the goat is tethered in a field to a rope which is 10 m long. In this field there is a rectangular shed of, for example, 2 m by 3 m. Get pupils to look at the locus of points that Percy can reach for different positions of the tethering post, relative to the shed. Here are some examples pupils could consider:

   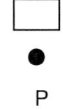

P   P   P   P

You might also introduce some two shed problems, such as:

Main teaching points

Loci problems essentially involve the same skills as defining a locus. However, the problems attempt to be practical and they are usually in several parts. Ensure pupils read the whole question and then use the skills developed in Section 15.3 to complete these (usually) more complex problems.

The starters in Sections 15.3 and 15.4 are possible GCSE questions. It is worth revisiting them to ensure understanding. Drawing circles (such as radio masts and/or the range of tethered animals) and bisecting lines and angles are examiners' favourites.

Common mistakes

Lack of accuracy; ±2 mm is the maximum error allowed in GCSE.

Plenary

Using a board pen with a piece of string (at least 50 cm long), get a series of volunteers to show how to do some of the questions from Exercise 15D of the Pupil Book (especially questions 1 and 5).

Similarity

16

Overview

16.1 Similar triangles
16.2 Areas and volumes of similar shapes

Section 16.1 shows pupils what similar figures are, how to recognise them, how to find the scale factor and how to use ratio to find corresponding lengths of similar figures. Section 16.2 builds on these ratio skills to include ratios using area and volume.

Context

This material has applications in problems involving length, area and volume of models, including anything from working out the heights of trees to making models of buildings.

AQA B references

AO3 Shape, space and measures: Geometrical reasoning

16.1 3.2 h "understand similarity of triangles and of other plane figures..."

AO3 Shape, space and measures: Transformations and coordinates

16.1–16.2 3.3 d "...identify the scale factor of an enlargement as the ratio of the lengths of any two corresponding line segments; ... understand and use the effect of enlargement on areas and volumes of shapes and solids"

Route mapping

Exercise	D	C	B	A	A*
A		1–5	6–8		
B			1–6	7	
C			all		
D				1–11	12–18
E					all

Answers to diagnostic Check-in test

1
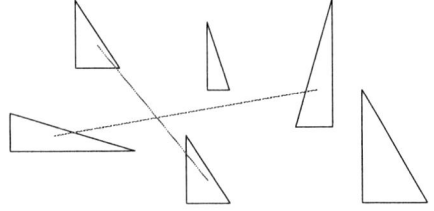

2 3 : 4

3 **a** 9 **b** 64 **c** 81

4 **a** 1 **b** 0.25

1 Link the pairs of congruent triangles with a line.

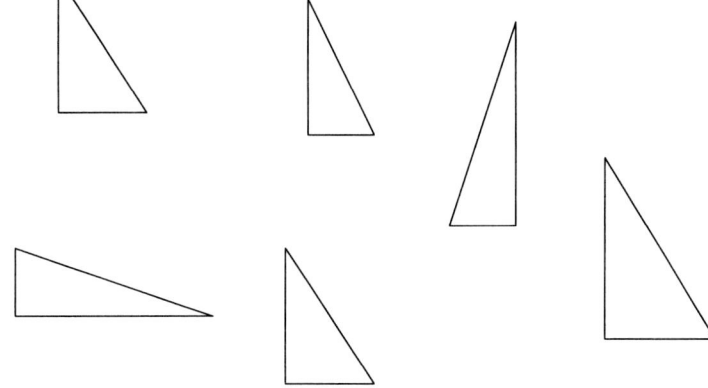

2 In his pencil case, Greg has 15 pencils and 20 coloured pens. What is the ratio of pencils to pens in its simplest form?

3 Write down the answer to:

a $3^2 =$ _____   **b** $4^3 =$ _____   **c** $9^2 =$ _____

4 Find the value of x. Show your method.

a $\dfrac{x}{8} = \dfrac{4}{32}$

b $\dfrac{x}{8} = \dfrac{4}{128}$

Incorporating exercises: 16A, 16B, 16C	**Key words**
Homework: 16.1	similar
Examples: 16.1	

Learning objective(s)

● show two triangles are similar
● work out the scale factor between similar triangles

Prior knowledge

Pupils should understand the meaning of congruency. They should know how to calculate a ratio (and how to cancel it down). Most important here is to be able to solve problems such as:

$$\frac{x}{3} = \frac{6}{9}$$

Starter

Draw three triangles (two congruent and one similar) on the board, like so:

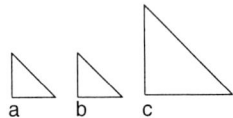

Ask pupils what the word that describes the link between a and b is? What about a and c? They are the same shape, but are they congruent? In this way, introduce the difference between similar and congruent triangles.

Provide some introductory ratio exercises. For example, if a class has 10 boys and 15 girls, ask pupils to write down the ratio of girls to boys and cancel it down. Then get them to solve equations such as:

$$\frac{x}{5} = \frac{8}{20} \quad \text{and} \quad \frac{x}{10} = \frac{15}{25}$$

Main teaching points

When making a ratio of sides, make sure that pupils are looking at corresponding sides. Sometimes redrawing helps comprehension. For example, some pupils might find it confusing to identify the corresponding sides in these triangles.

Redrawing and realigning the triangles (as below) makes the problem much easier.

Similarly, questions that involve similar triangles with shared (or overlapping) sides can be easier if the two triangles are drawn out separately.

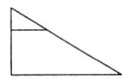　　　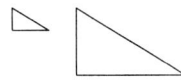

Common mistakes

The most frequent mistake is for pupils to not use corresponding sides when making a ratio.

Differentiation

Help lower ability pupils by insisting they redraw shapes in the same orientation and draw arrows from one side to the corresponding side.

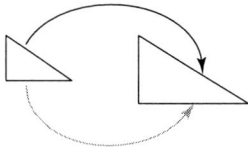

Plenary

Draw similar triangles on the board, but in different orientations. For example:

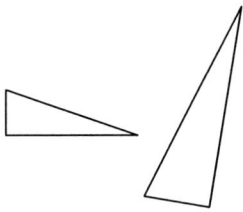

Ask pupils to give you different information to generate different types of questions. For example, all three side lengths of one triangle, and one side length of the other.

How would each question be solved? (Remember to start by re-drawing the triangles.)

Do the same for questions with triangles such as:

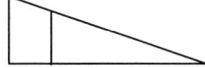

			Key words	
Incorporating exercises:	16D, 16E		area ratio	volume ratio
Homework:	16.2		area scale factor	volume scale
Examples:	16.2		length ratio	factor
			linear scale	
			factor	

Learning objective(s)

- solve problems involving the area and volume of similar shapes

Prior knowledge

Pupils will need to know the squares and cubes of integers and be successful in completing Exercises 16A, 16B and 16C.

Starter

Get pupils to draw a line 3 cm long. Now ask them to draw another line twice as long. What is the length of this line?

Similarly draw a square 3 cm by 3 cm. What is the area of this square? Draw another square with sides twice as long. What is the area of this square? Ask what has happened to the area: it is not twice as big, why?

Ask pupils to consider what would happen to the volume if they drew cubes of side length 3 cm and 6 cm?

Main teaching points

The key teaching point is to get pupils to understand the length, area and volume ratios. Regardless of the shape, if two shapes are similar and their lengths are in the ratio $x : y$, then the ratio of their areas will always be $x^2 : y^2$, and the ratios of their volumes will always be $x^3 : y^3$.

Common mistakes

The most common mistake pupils make is to find the areas and volumes using the same ratio as the lengths. For example, if the ratio of lengths is 1 : 2, some pupils may use this to calculate the area instead of using 1 : 4, or use it to find the volume instead of using 1 : 8.

Plenary

Ask pupils to work out the ratios of the areas of these squares. (**Answers** 1 : 4, 1 : 100, 1 : 25)

Suggest they then repeat the exercise for the ratios of the volumes of cubes with sides as shown below.

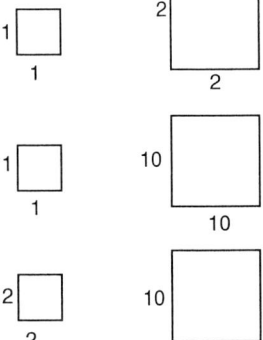

Dimensional analysis

This chapter covers the dimensional analysis required at higher level, looking at examples of formulae for length, area and volume, leading to the main idea of consistency of dimension.

AQA B references

AO3 Shape, space and measures: Transformations and coordinates

17.1–17.4   3.3d "… understand the difference between formulae for perimeter, area and volume by considering dimensions; …"

Route mapping

Exercise	D	C	B	A	A*
A	1–8	9–10			
B	1–5	6–10			
C	1	2–4	5–6		
D		1–3	4–6		

Answers to diagnostic Check-in test

a lw

b $2l + 2w$

c $2\pi r$

d $\dfrac{\pi d^2}{4}$

e xyz

f $2xy + 2yz + 2zx$

g $\dfrac{4}{3}\pi r^3$

h $4\pi r^2$

i πdh

j $\dfrac{3\pi d^2}{4}$

1 Write down a formula for:

a the area of a rectangle with dimensions l and w.

b the perimeter of a rectangle with dimensions l and w.

c the circumference of a circle with radius r.

d the area of a circle with diameter d.

e the volume of a cuboid with dimensions x, y and z.

f the surface area of a cuboid with dimensions x, y and z.

g the volume of a sphere of radius r.

h the surface area of a sphere of radius r.

i the curved surface area of a cylinder of diameter d and height h.

j the total surface area of a hemisphere of diameter d.

17.1 Dimensions of length

Incorporating exercise:	17A
Homework:	17.1
Example:	17.1

Key words
1-D
length
perimeter

Learning objective(s)

● find formulae for the perimeter of 2-D shapes

Prior knowledge

Pupils should know the formulae for perimeters of various shapes, and the metric units commonly used to express length: mm, cm, m, km.

Starter

Ask pupils for perimeter formulae for various shapes. For example, suggest to them a circle of radius r, a rectangle of dimensions l by w, a regular pentagon of side x and an equilateral triangle of side y. Ask them to give expressions for the perimeters.

Main teaching points

The main teaching point is the difference between lengths (having one dimension), and numbers (which are dimensionless). This might be illustrated by using the fact that length + length = length, while number × length = length. (Although number + length is undefined, the idea of consistency is left until Section 17.4.)

Exercise 17A in the Pupil Book simply focuses on finding expressions for the perimeter of various polygons. Finding expressions for the perimeters of various shapes should be relatively straightforward at higher level. Some difficulties may arise from simplifying the algebraic expression obtained for the perimeter, but these will usually be simple and should not cause too many problems.

Plenary

Revisit the answers obtained from either Exercise 17A or the homework sheet. Point out that each perimeter can be seen to be either:
a a length
b a number × length (= length)
c a sum of the above (= length)

			Key words
Incorporating exercise:	17B		2-D
Homework:	17.2		area
Example:	17.2		

Learning objective(s)

● find formulae for the area of 2-D shapes

Prior knowledge

Pupils should know the formulae for areas of standard shapes.

Starter

Ask pupils for area formulae for various shapes. Suggest to them a circle of radius r, a rectangle of dimensions l by w, an equilateral triangle of side y, and a regular hexagon of side h. Ask them to calculate expressions for areas.

Main teaching points

The main teaching point is the difference between areas (having two dimensions), lengths (having one dimension) and numbers (which are dimensionless).

This might be illustrated by using the fact that area = length × length, while number × area = area. The focus in Exercise 17B in the Pupil Book is simply on finding algebraic expressions for the areas of various compound shapes. This has been covered before in Chapter 11 and so should provide few difficulties for pupils.

Plenary

Revisit the answers obtained from either Exercise 17B or the homework sheet. Point out that each area can be seen to be either:
a a length × length
b a number × length × length (= area)
c a sum of the above (= area)

Incorporating exercise:	17C	Key words
Homework:	17.3	3-D
Example:	17.3	volume

Learning objective(s)

● find formulae for the volume of 3-D shapes

Prior knowledge

Pupils should know how to find the volumes of simple and compound solids, including prisms.

Starter

Ask pupils to recall the volume formulae for standard shapes: a sphere of radius r, a cube of side x, a cylinder of radius r and height h, and a general prism of uniform cross-section.

Main teaching points

The main teaching point is the difference between volumes (having three dimensions), areas (having two dimensions), lengths (having one dimension) and numbers (which are dimensionless).

The dimensional aspects of volume should be illustrated by focusing on these examples:

● the fact that volume = area × length can be seen from the general formula for the volume of a prism
● volume = length × length × length could be illustrated by looking at the volume of a cube or cuboid
● number × volume = volume can be seen by taking a cuboid of lengths l, w and h, calculating its volume, then changing the length of w to $2w$, for example.

The focus of Exercise 17C is simply on finding algebraic expressions for the volumes of various shapes.

Plenary

Revisit the answers obtained from either exercise 17C in the Pupil Book or the homework sheet. Point out that each volume can be seen to be either:
a a (number ×) area × length
b a (number ×) length × length × length
c a sum of the above (see, for example, numbers 3 and 5 from Homework 17.3).

Incorporating exercise:	17D	**Key words**
Homework:	17.4	consistency
Examples:	17.4	dimension
		formula

Learning objective(s)

● check that the dimensions of a formula are consistent

Prior knowledge

Pupils should know how to calculate the perimeters, areas and volumes of various shapes as covered in Sections 17.1–17.3.

Starter

Say to the class: "It is suggested that the formula for the volume of a hemisphere is $\frac{2}{3}\pi r^2$."

Ask them why this formula *must* be incorrect. (Because number × length × length = area.)

Repeat the exercise for a made-up shape. Say the curved surface area of an object is $\frac{1}{2}\pi r^2 + \frac{4}{5}\pi r^2 h$.

Ask them again why it must be incorrect. (Because area + volume is inconsistent.)

Main teaching points

A formula's consistency may be checked by ensuring that every term in the formula has the same dimension. Note that this does not necessarily mean the formula will be correct! For example, it might be suggested that the formula for the volume of a cone is:

$$\frac{2}{3}\pi r^2 h$$

Although it is dimensionally correct ($r \times r \times h$ = length × length × length = volume), we know from experience that the formula itself is wrong. Checking for consistency is therefore one possible way of deciding if a formula is definitely wrong. However, it can never tell you that a formula is correct, only that it is possibly so.

Plenary

Summarise the dimension notes for pupils as used throughout this chapter.
● number × length = length
● length + length = length
● length × length = area
● area + area = area
● length × length × length = volume
● area × length = volume
● volume + volume = volume
● volume + area is inconsistent
● area + length is inconsistent

    Module 5: Algebra and Space, shape and measure

Overview

18.1 Properties of vectors
18.2 Vectors in geometry

This chapter builds on pupils' knowledge of vectors as a means of describing translations. Section 18.1 covers how to add and subtract vectors as well as the simple multiplication of vectors on a grid. Section 18.2 covers how to use vectors to solve geometrical problems.

Context

Vectors can be used in calculations involving velocity, acceleration and various forces.

AQA B references

AO3 Shape, space and measures: Transformations and coordinates

18.1–18.2 3.3f "understand and use vector notation; calculate, and represent graphically the sum of two vectors, the difference of two vectors and a scalar multiple of a vector; calculate the resultant of two vectors; understand and use the commutative and associative properties of vector addition; solve simple geometrical problems in 2-D using vector methods"

Route mapping

Exercise	C	B	A	A*
A			1–10	11–15
B				all

Answers to diagnostic Check-in test

1
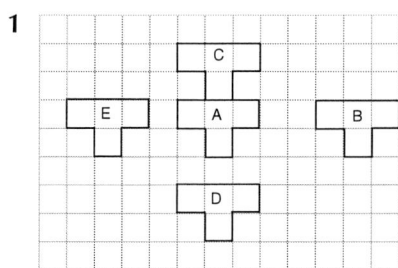

2 a $\begin{pmatrix} 3 \\ 0 \end{pmatrix}$ b $\begin{pmatrix} 5 \\ 1 \end{pmatrix}$ c $\begin{pmatrix} 1 \\ 7 \end{pmatrix}$ d $\begin{pmatrix} -1 \\ 4 \end{pmatrix}$ e $\begin{pmatrix} 1 \\ -4 \end{pmatrix}$ f $\begin{pmatrix} 6 \\ 0 \end{pmatrix}$ g $\begin{pmatrix} 2 \\ 3 \end{pmatrix}$ h $\begin{pmatrix} -2 \\ -3 \end{pmatrix}$ i $\begin{pmatrix} 4 \\ -3 \end{pmatrix}$ j $\begin{pmatrix} 3 \\ 0 \end{pmatrix}$ k $\begin{pmatrix} -3 \\ 0 \end{pmatrix}$ l $\begin{pmatrix} 0 \\ 3 \end{pmatrix}$

1 For each part of this question, always start with shape A. Draw the image of shape A after translating it as follows.

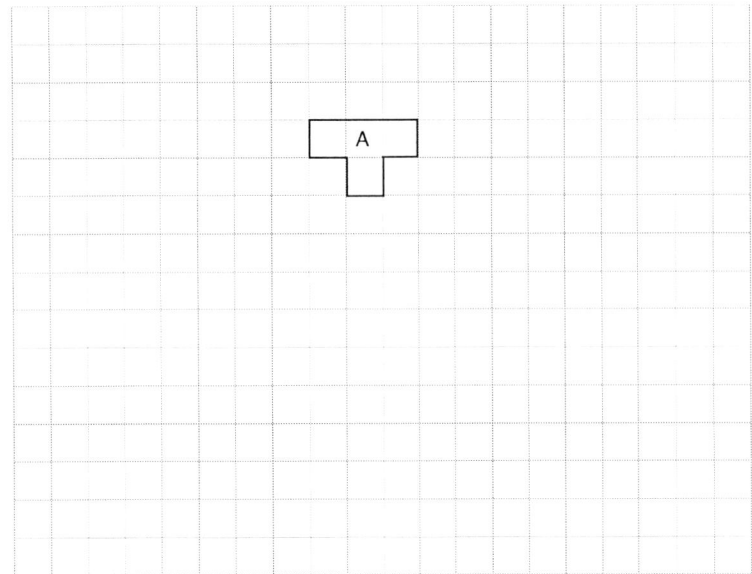

a Translate 5 squares right. Label this shape B.
b Translate 2 squares up. Label this shape C.
c Translate 3 squares down. Label this shape D.
d Translate 4 squares left. Label this shape E.

2 Write down the vector which describes each of the following translations.

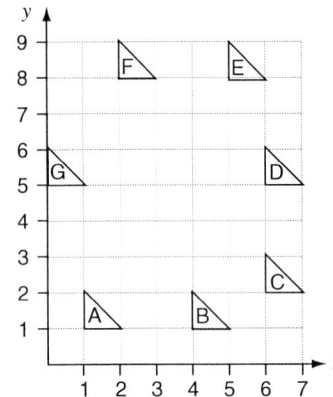

a A to B _____

b A to C _____

c A to F _____

d A to G _____

e G to A _____

f G to D _____

g G to F _____

h F to G _____

i F to D _____

j F to E _____

k E to F _____

l C to D _____

Module 5: Algebra and Space, shape and measure

		Key words
Incorporating exercise:	18A	direction
Homework:	18.1	magnitude
Example:	18.1	vector

Learning objective(s)

● add and subtract vectors

Prior knowledge

Pupils must be able to use column vectors to describe translations.

Starter

Draw these points on a square grid.

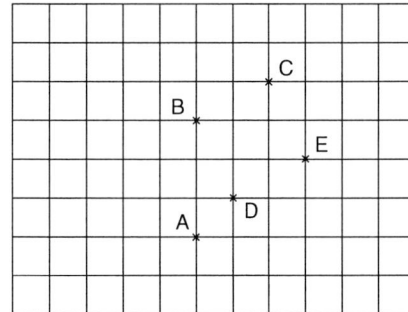

Get the pupils to give the column vectors for $\overrightarrow{AB}$, $\overrightarrow{BA}$, $\overrightarrow{CE}$, $\overrightarrow{EC}$, etc.

Using the same grid, explain that one unit across = **a** and one unit up = **b**. Get the pupils to express the column

vectors for $\overrightarrow{AB}$, $\overrightarrow{BA}$, $\overrightarrow{CE}$, $\overrightarrow{EC}$, etc. in terms of **a** and **b**.

Main teaching points

Make sure that pupils are familiar and comfortable with the representation and notation used for vectors. They should be clear that a vector has both magnitude and direction. For example, this diagram represents a vector:

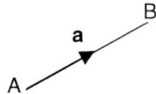

The direction of the vector is shown by an arrow, usually in the middle of the line as shown. The magnitude of the vector is the length of the line. This vector can be represented by **a** (always in bold), or <u>a</u>, or $\overrightarrow{AB}$.

There are no clear rules about which of the three representations of a vector is used, so pupils need to be told to expect to see either: **a**, <u>a</u> or $\overrightarrow{AB}$. In the Pupil Book, and in the GCSE exams, pupils will not see the <u>a</u> type but this may be used in some books.

If the vector from A to B, $\overrightarrow{AB}$, is given a value of **a**, then the vector from B to A, $\overrightarrow{BA}$, has the value of –**a**.

Differentiation

Most good grade B pupils should be able to tackle this topic – especially when there are grids involved. You should encourage pupils to draw on the grids if it helps them complete the questions. Many pupils will, however, struggle on questions 12–15 of Exercise 18A in the Pupil Book.

Plenary

On a grid on the board draw a vector $\overrightarrow{AB}$. Ask someone to give the column vector for $\overrightarrow{AB}$.

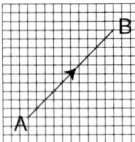

Now add point C, drawing the vectors $\overrightarrow{AC}$ and $\overrightarrow{CB}$. Ask what $\overrightarrow{AC} + \overrightarrow{CB}$ equals, and why.

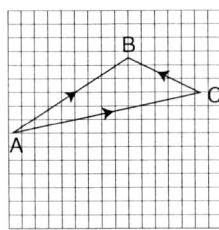

Returning to your original vector diagram $\overrightarrow{AB}$, add new points C, D, E, F and G. Again, ask what would be the sum of all the individual vectors $\overrightarrow{AC}$, $\overrightarrow{CD}$, $\overrightarrow{DE}$, $\overrightarrow{EF}$, $\overrightarrow{FG}$, $\overrightarrow{GB}$.

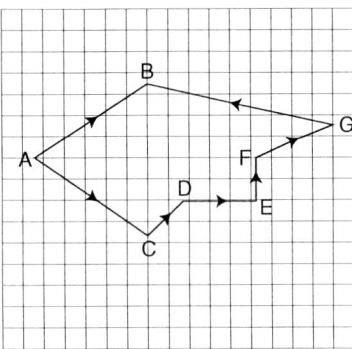

			Key words
Incorporating exercise:	18B		vector
Homework:	18.2		
Example:	18.2		

Learning objective(s)

- use vectors to solve geometrical problems

Prior knowledge

Pupils should be confident finding and using vectors such as $\frac{3}{2}\mathbf{a} - \frac{3}{2}\mathbf{b}$ and $2(2\mathbf{a} - \mathbf{b}) - 3(\mathbf{a} - \mathbf{b})$.

Starter

Draw the parallelogram PQRS. Ask pupils to express $\overrightarrow{PS}$, $\overrightarrow{PQ}$, $\overrightarrow{PR}$ and $\overrightarrow{QS}$ in terms of $\mathbf{a}$ and $\mathbf{b}$. (Answers: $\overrightarrow{PS} = \mathbf{a}$, $\overrightarrow{PQ} = \mathbf{b}$, $\overrightarrow{PR} = \mathbf{a} + \mathbf{b}$, $\overrightarrow{QS} = \mathbf{a} - \mathbf{b}$)

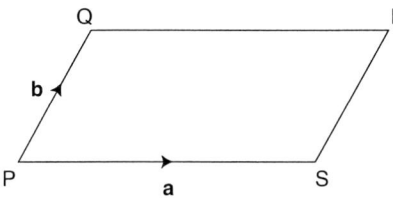

Draw the triangle PQS. What is $\overrightarrow{QS}$ in terms of $\mathbf{a}$ and $\mathbf{b}$? (Answer: to get from Q to S you need to go from Q to P then P to S, so $\overrightarrow{QS} = -\mathbf{b} + \mathbf{a}$ (or $\mathbf{a} - \mathbf{b}$))

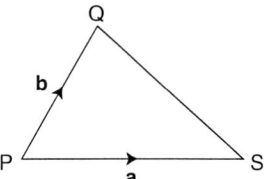

Main teaching points

Those pupils who found Section 18.1 easy will probably cope. Reading questions twice is a prerequisite for success in this topic. Many pupils make mistakes by trying to rush their answers.

Remind pupils that questions may require Pythagoras and occasionally trigonometry.

Differentiation

Only your best pupils need apply! This is a true A* topic.

Plenary

Talk through how one could find the vector $\overrightarrow{AM}$, where M is the midpoint of DE.

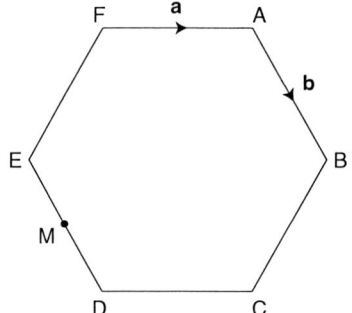

Algebra 1

Overview

Sections 19.1 to 19.3 are aimed at building upon the pupils' knowledge of algebraic manipulation. Section 19.4 is that grade C favourite, trial and improvement. Section 19.5 is concerned with the solving and setting up of simultaneous equations. The chapter then finishes with transposing formulae in Section 19.6.

Context

Solving equations, dealing effectively with algebraic expressions, and re-arranging formulae is the basis of much of maths and science, and of many other subjects ranging from economics to architecture.

AQA B references

AO2 Number and algebra: Calculations

19.3, 19.6 2.3a "... use inverse operations, understanding that the inverse operation of raising a positive number to power n is raising the result of this operation to power (1 divided by n)"

19.2, 19.3, 19.6 2.3b "use brackets and the hierarchy of operations"

AO2 Number and algebra: Equations, formulae and identities

19.1–19.6 2.5a "distinguish the different roles played by letter symbols in algebra, using the correct notational conventions for multiplying or dividing by a given number, and knowing that letter symbols represent definite unknown numbers in equations, defined quantities or variables in formulae, general, unspecified and independent numbers in identities, ..."
2.5b "understand that the transformation of algebraic entities obeys and generalises the well-defined rules of generalised arithmetic ... manipulate algebraic expressions by collecting like terms, multiplying a single term over a bracket, taking out common factors ..."
2.5c "know the meaning of and use the words 'equation', 'formula', ... and 'expression'"
2.5d "use index notation for simple integer powers, and simple instances of index laws; substitute positive and negative numbers into expressions such as $3x^2 + 4$ and $2x^3$"

19.3	2.5f "set up simple equations; solve simple equations by using inverse operations or by transforming both sides in the same way"				

19.3 2.5f "set up simple equations; solve simple equations by using inverse operations or by transforming both sides in the same way"
2.5g "solve linear equations in one unknown, with integer or fractional coefficients, in which the unknown appears on either side or on both sides of the equation; solve linear equations that require prior simplification of brackets, including those that have negative signs occurring anywhere in the equation, and those with a negative solution"

19.6 2.5h "… change the subject of a formula including cases where the subject occurs twice, or where a power of the subject appears …"

19.5 2.5j "find the exact solutions of two simultaneous equations in two unknowns by eliminating a variable …"

19.4 2.5m "use systematic trial and improvement to find approximate solutions of equations where there is no simple analytical method of solving them"

Route mapping

Exercise	D	C	B	A	A*
A	all				
B	all				
C	1	2–6			
D	1–16	17–22			
E		all			
F	all				
G	1–8	9–14			
H	1–4	5–8			
I		all			
J			all		
K			all		
L			all		
M			1–3	4–9	
N		1–12	13–21		

Answers to diagnostic Check-in test

1 14 **2** −4 **3** 20 **4** 15 **5** $6x$

6 $12x$ **7** $10 + 6x$ **8** $6a$ **9** $8p^2$ **10** a^4

1 $2 + 3 \times 4 =$

2 $-7 - (-3) =$

3 $4(2 + 3) =$

4 $5(4 - 1) =$

5 $x(4 + 2) =$

6 $3x(6 - 2) =$

7 $10 + (2x + 4x) =$

8 $3 \times 2a =$

9 $2p \times 4p =$

10 $a^3 \times a =$

Incorporating exercise: 19A **Homework:** 19.1a 💿 **Examples:** 19.1a	**Key words** bracket expression coefficient substitute like terms simplification expand variable

Learning objective(s)

● substitute into, manipulate and simplify algebraic expressions

Prior knowledge

Pupils should be conversant with the basic language of algebra, BODMAS, and negative numbers.

Starter

Draw this grid on the board without the answers filled in:

		a		
		2	3	5
	2	*6*	*8*	*12*
b	4	*8*	*10*	*14*
	6	*10*	*12*	*16*

Get the pupils to fill in the values by using a formula, for example $2a + b$.

Explain, if needed, that $2a + b$ is 2 times a, plus b.

Give a time limit or make it a race.

Draw the same empty grid again, but give a different rule.

Main teaching points

Take time to emphasise the role of BODMAS once substitution has taken place.

Common mistakes

Ignoring BODMAS rules.

Misunderstanding algebraic conventions. For the expression $2t + 7$ where $t = 4$, a common error is to rewrite the expression as $24 + 7$ instead of $2 \times 4 + 7$.

Differentiation

Although all questions are grade D, concentrate on questions like 2, 3, and 9, with whole-number values for substitution, for the lower ability pupils.

Plenary

Ask pupils to find the values of some expressions, such as $50 - 3p$ (when $p = -4$), $5t^2$ (when $t = 3$ and when $t = -3$), $(\frac{1}{q^2})$ (when $q = 2$) etc., to emphasise the use of BODMAS and some of the difficulties encountered with a negative sign.

Incorporating exercise:	19B		**Key words**	
Homework:	19.1b		bracket	expression
Examples:	19.1b		coefficient	substitute
			like terms	simplification
			expand	variable

Learning objective(s)

● substitute into, manipulate and simplify algebraic expressions

Prior knowledge

Pupils need to be able to multiply algebraic terms together.

Starter

Ask a variety of multiplication questions, for example 5×2, $5 \times 2a$, $5 \times 2a^2$, $5a \times 2a$, $5a \times 2b$, $5a \times 2a^2$, $5a^2 \times 2a$, $5a \times 2a^3$, etc.

Main teaching points

Brackets: a method of multiplying one term outside the bracket, by one *or more* terms inside the bracket. Emphasise that *everything* inside the bracket is multiplied by the number or algebraic term outside the bracket.

Common mistakes

Multiplying the first term properly but then just writing down the subsequent terms without multiplying them. For example, $5(3a + 2b) = 15a + 2b$ (it should be $15a + 10b$).

Differentiation

For low ability pupils, encourage the drawing of 'arcs' as the multiplying is done, to keep track of what has been done. For example: $5(3a + 2b) = 15a + 10b$.

Also, give plenty of help when powers are involved, especially when the expression involves the power of 3, or 4, or more.

Plenary

● Expand: ☺(✳+ ☺)
● Expand: 2☺(2♥ −3☺)

Pupils think this is a bit of silly fun, but it is essential that they grasp the idea that:
     ☺(✳+ ☺) = ☺✳+ ☺² and 2☺(2♥ − 3☺) = 4☺♥ −6☺²

Incorporating exercise:	19C	Key words	
Homework:	19.1c	bracket	expression
Examples:	19.1c	coefficient	substitute
		like terms	simplification
		expand	variable

Learning objective(s)

- substitute into, manipulate and simplify algebraic expressions

Prior knowledge

Pupils should have completed Section 19.1b Basic algebra: Expansion.

Starter

Ask the following quick-fire questions:
- What is 3 apples + 4 apples? (7 apples)
- What is $3a + 4a$? ($7a$)
- What is 3 apples + 4 apples + 2 bananas + 3 bananas? (7 apples + 5 bananas)
- What is $3a + 4a + 2b + 3b$? ($7a + 5b$)
- Expand $4(3 + 5a)$. ($12 + 20a$)
- Expand $3(3a - 2)$. ($9a - 6$)

Main teaching points

Take time to identify like terms, and remember you cannot add unlike terms together.

Common mistakes

Combining unlike terms, for example $4a + 5 = 9a$. (The correct answer is that this cannot be simplified.)

Differentiation

With lower ability pupils, break the 'expand and simplify' questions into two sections. Expand first (mark that) and then simplify (mark again).

Plenary

These examples will help pupils to focus on the process of simplification, and hopefully will clear up any misunderstandings that the weaker pupils may have.
- Simplify $3a + 4b + 5a$ ($8a + 4b$)   Why is the answer not $12ab$? Discuss.
- Simplify $2t + 3t^2$ (you can't)   Why is the answer not $5t^2$ or $5t^3$? Discuss.

Incorporating exercise:	19D	**Key words**
Homework:	19.2	common factor
Examples:	19.2	factorisation

Learning objective(s)

● factorise an algebraic expression

Prior knowledge

Being able to expand brackets is the first step to being able to factorise, because one is the inverse of the other. Pupils should be able to find the HCF (highest common factor) of two or more numbers.

Starter

What is the HCF of each of the following?

6 and 9? (3) $6a$ and 9? (3) $6a$ and $9a$? ($3a$)

$6a^2$ and $9a$? ($3a$) $6a^2$ and $9a^2$? ($3a^2$)

Main teaching points

One can always check an expression has been factorised correctly by expanding the answer. When factorising, pupils must look for the *HCF* and not just *any* factor.

Common mistakes

Finding *a* factor, and not the *highest* factor.

Differentiation

For low achievers, from Exercise 19D, do questions 1 to 5 and others like them, before attempting questions 6 onwards. Rewrite question 6 as $5 \times g \times g + 3 \times g$ to help explain the $5g^2$ part before they go further.

Plenary

Ask the pupils to supply an expression with terms that have an HCF of 6, $6a$, $6a^2$, $6ab$, $6ab^2$, etc. Ask for several expressions for each HCF. (This is the reverse of the starter.)

		Key words	
Incorporating exercise:	19E	brackets	rearrange
Homework:	19.3a	do the same to	solution
Example:	19.3a	both sides	solve
		equation	

Learning objective(s)

● solve equations in which the variable appears as part of the numerator of a fraction

Prior knowledge

Pupils need to be able to solve simple one- and two-step equations where the variable appears only on one side of the equation, for example $2x - 4 = 6$.

Starter

Ask pupils to solve the following equations.

$x - 4 = 6$

$2x - 4 = 6$

$2x + 4 = 6$

$\frac{x}{3} = 6$

Main teaching points

Following the simple algebraic rules of manipulation means that pupils should cope with these questions efficiently after a few good examples.

Common mistakes

The main mistake made with fractional equations is that the pupils will try to solve the problem 'out of order'.

For example, with $\frac{x + 5}{3} = 7$, pupils will often try to $- 5$ rather than $\times 3$ as the first step, as they don't realise

that the $x + 5$ must remain together until the denominator is removed.

Differentiation

Using brackets around the numerator may help some pupils understand and/or remember that the denominator must be removed before the numerator is dealt with (if it is more complex than just x or $3y$ etc.).

Plenary

Get help from the class to solve these three equations, and ask the pupils to point out things to watch out for. For example: How might they be confused? What mistakes might they make?

$$x + \frac{10}{2} = 20 \qquad \frac{x}{2} + 10 = 20 \qquad \frac{x + 10}{2} = 20$$

Incorporating exercise:	19F
Homework:	19.3b
Example:	19.3b

Key words

brackets   rearrange
do the same to   solution
  both sides   solve
equation

Learning objective(s)

● solve equations where you have to expand brackets first

Prior knowledge

Pupils must know how to expand a bracket.

Starter

Ask the following quick-fire questions:
● What is $3 \times £4.25$? (£12.75)

  What is $3 \times £4 + 3 \times 25p$? ($£12 + 75p = £12.75$)

  What is $3 \times (£4 + 25p)$? ($£12 + 75p = £12.75$)

  What is $3(£4 + 25p)$? ($£12 + 75p = £12.75$)

● Is $5(2 + 4p) = £10.20$? (No $= 10 + 20p$)

● What is $5(2x + 3y)$? ($10x + 15y$)

  Emphasise that the answer is $5 \times 2x + 5 \times 3y$.

Main teaching points

Brackets have to be multiplied out before equations can be solved. Everything inside the bracket must be multiplied by what is outside. Care must be taken when dealing with negative numbers. Once the brackets are removed the equations can be solved as before.

Common mistakes

Only multiplying the first part of the inside of the brackets by the outside.

Plenary

Check all pupils can correctly multiply out brackets by individual questioning.

Solving linear equations: Equations with the variable on both sides

Incorporating exercise:	19G	Key words	
Homework:	19.3c	brackets	rearrange
Example:	19.3c	do the same to both sides	solution
		equation	solve

Learning objective(s)

- solve equations where the variable appears on both sides of the equation

Prior knowledge

Pupils should know how to expand a bracket.

Starter

Solve the algebra pyramids below. Add together the bricks on the bottom row to get the middle row. The top row is the solution to the equation. Solve to find x.

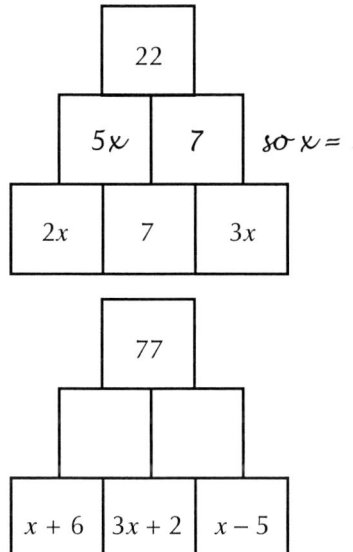

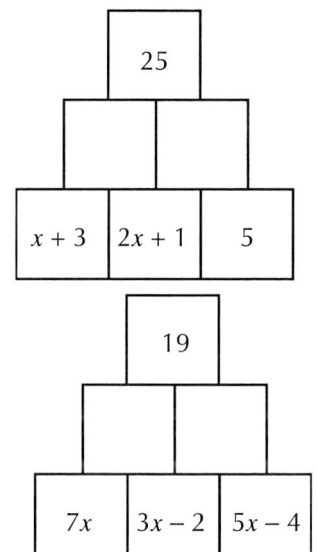

Main teaching points

It is usually easiest to use the 'do the same to both sides' method when solving these equations. It is necessary to get the variable to one side of the equation and the constant to the other side. It is usual to collect the variable on the left hand side but when this means pupils will end up with a negative variable it is more sensible to turn the equation round or collect on the right hand side.

Common mistakes

Not changing the signs when changing sides.

Differentiation

In Exercise 19G, questions 1–8 are grade D, 9–14 are grade C.

Plenary

Check for understanding by going through some of the later questions.

Module 5: Algebra and Space, shape and measure   © HarperCollins*Publishers* Ltd 2006   167

Incorporating exercise:	19H
Homework:	19.3d
Example:	19.3d

Key words

brackets	rearrange
do the same to	solution
both sides	solve
equation	

Learning objective(s)

● set up equations from given information, and then solve them

Prior knowledge

Pupils must be able to solve simple equations.

Starter

Ask the pupils to formulate equations for simple word problems such as, "If you cut m cm from 2 metres of ribbon, how much is left?" "The side of a square is c metres. What is its area?"

Ask the pupils to give you any simple formulae they know. They may give these to you in words which, as a class, you can convert into algebra.

Main teaching points

This section covers using equations in real life to solve problems. Pupils need to focus on what is important mathematically in solving these problems. It is often simplest to use letters that relate to the questions, hence b for the number of bottles in Example 14. Stress to your pupils that the letter stands for a number and not an item, that is, the *number* of bottles in the crate. Pupils should check their answers and be careful to use the correct units.

Differentiation

In Exercise 19H, questions 1–4 are grade D, 5–8 are grade C.

Plenary

Ask the pupils to give you simple problems, and provide the equations.

19.4 — Trial and improvement

|---|---|
Incorporating exercise: 19I	**Key words**
Homework: 19.4	comment trial and
Example: 19.4	decimal place improvement
	guess

Learning objective

- estimate the answer to some equations that do not have exact solutions, using the method of trial and improvement

Prior knowledge

Pupils should understand approximation and estimation.

Starter

Put the pupils into pairs. One thinks of a number between 1 and 100 and writes it down without the other seeing it. The partner has 10 goes to guess the number. Roles are reversed. See who discovered the number in the least number of guesses.

Main teaching points

Not all equations can be solved exactly and so a mechanism needs to be found to find an approximate solution. Trial and improvement is a way of doing this. Emphasise that it is trial and improvement not trial and error. These questions appear on the calculator paper, pupils need to understand that they must show their working and their solutions in order to gain the method marks.

Common mistakes

Pupils do not put the value of the variable as their answer, but the value of the equation.

Plenary

Discuss question 7 of Exercise 19I.

Incorporating exercises:	19J, 19K, 19L
Homework:	19.5a
Examples:	19.5a

Key words

balance	simultaneous
check	equations
coefficient	substitute
eliminate	variable

Learning objective(s)

● solve simultaneous linear equations in two variables

Prior knowledge

Pupils must be able to rearrange and solve basic equations such as $4x - 6 = 2$.

Starter

Make sure the pupils have their books closed.
Ask "If $x + y = 10$, what is the value of x?"
Assuming someone ventures a number, say 5, ask, "What is the value of y when $x = 5$? Write down some of the pairs of solutions; make sure $x = 4$ and $y = 6$ are included. Get the pupils to realise that there are many (indeed infinite) values of both x and y.
Ask the same of $2x + y = 14$. Write down possible values of x and y; make sure $x = 4$ and $y = 6$ are included.
Point out that $x = 4$ and $y = 6$ satisfy both of the equations at the same time, that is, simultaneously.
Some pupils might appreciate a quick sketch of the two graphs, showing that they intersect at the point (4, 6).

Main teaching points

Always try to use the elimination method in favour of the substitution method, as this is the main GCSE type question.

Common mistakes

Getting in a muddle with negatives, for example, when subtracting the two equations, and one of the numbers is already a negative.

Differentiation

With weaker pupils, it may be worth demonstrating the elimination method only.

Plenary

Put Example 18 (page 420) and then Example 19 (page 421) on the board and get the pupils to talk through them – ask "Why?" as often as possible.

Incorporating exercise:	19M	
Homework:	19.5b	
Example:	19.5b	

Key words

balance	simultaneous
check	equations
coefficient	substitute
eliminate	variable

Learning objective(s)

● solve simultaneous linear equations in two variables

Prior knowledge

Pupils need to be able to solve simultaneous equations.

Starter

Ask the class to solve the following simultaneous equations:
$5x - 2y = 26$
$3x - y = 15$
$(x = 4, y = -3)$

Main teaching points

Setting up simultaneous equations involves reading the question very carefully, then translating the English question into a Maths question. Use whatever letters you like, but make them obvious (it is usually made easy by the question), and if in doubt, use x and y.

Common mistakes

Using x and y in both equations, but mixing up the meanings between the two equations.

Plenary

Get the pupils to make up their own questions for the following simultaneous equations:
a $4A + B = 17$ **b** $2C + 5D = 15$
   $2A + B = 9$ $3C - 2D = 13$

Incorporating exercise:	19N
Homework:	19.6
Example:	19.6

Key words

expression	transpose
rearrange	variable
subject	

Learning objective(s)

● rearrange formulae using the same methods as for solving equations

Prior knowledge

Pupils need to have the skills to be able to solve basic algebraic equations.

Starter

Ask the following questions:
● Solve the equation (that is, make x the subject): $2x - 4 = 6$
● Make x the subject of: $2x - y = 6$
● Make x the subject of: $2x - y = z$
● Make x the subject of: $wx - y = z$

Emphasise that all four equations are essentially the same.

Main teaching points

Letters are just numbers that we don't know the value of – there is no real difference between $2x - 4 = 6$ and $wx - y = z$.

Differentiation

In Exercise 19N, questions 1 to 12 are grade C, and the rest are grade B. If a pupil can complete most of the grade C questions, the grade B ones shouldn't be a problem as long as they are aware that the inverse of 'square' is 'square root'.

Plenary

Make y the subject of $x = y^2 + 5$. Ask the pupils to write down an incorrect solution, then get a volunteer to write their incorrect solution on the board – ask the rest of the class to find the mistake.

Overview

20.1 Expanding brackets
20.2 Quadratic factorisation
20.3 Solving quadratic equations by factorisation
20.4 Solving a quadratic equation by the quadratic formula
20.5 Solving a quadratic equation by completing the square
20.6 Problems using quadratic equations

This chapter deals with the various methods of factorising and solving quadratic equations, after a gentle warm-up of expanding brackets to make quadratic expressions.

Context

When the skills of this chapter have been mastered, then the last exercise shows a wide variety of different problems that can be solved using quadratic equations.

AQA B references

AO2 Number and algebra: Using and applying number and algebra

20.2–20.6 2.1a "select and use appropriate and efficient techniques and strategies to solve problems of increasing complexity, involving numerical and algebraic manipulation"

AO2 Number and algebra: Calculations

20.4, 20.5, 20.6 2.3q "use surds ... in exact calculations ..."
20.4, 20.6 2.3r "use calculators effectively and efficiently, knowing how to enter complex calculations; ..."

AO2 Numbers and Algebra: Equations, formulae and identities

20.2–20.6 2.5b "... factorising quadratic expressions including the difference of two squares and cancelling common factors in rational expressions"
20.3, 20.4, 20.5 2.5k "solve quadratic equations by factorisation, completing the square and using the quadratic formula"

Route mapping

Exercise	D	C	B	A	A*
A		all			
B			all		
C			all		
D		1–8	9–24		
E			all		
F			1–9	10–18	
G				all	
H		1–12	13–27	28–36	
I				all	
J				all	
K				1–2	3–5
L				all	
M				all	
N				all	

Answers to diagnostic Check-in test

1 a $7x + 16$ **b** $13x^2 + 4x$ **c** $14x^2 - 9x$

2 a $6(x + 2)$ **b** $3x(x + 3)$ **c** $4x(3y - x)$

3 a $12t^2$ **b** $-12p^2$

4 a $x = -\frac{7}{3}$ **b** $x = \frac{4}{5}$

      Module 5: Algebra and Space, shape and measure

1 Multiply out then simplify.

a $2(x + 3) + 5(2 + x)$

b $3x(x + 3) + 5x(2x - 1)$

c $4x(x - 1) - 5x(1 - 2x)$

2 Factorise the following expressions.

a $6x + 12$

b $3x^2 + 9x$

c $12xy - 4x^2$

3 Simplify the following.

a $6t \times 2t$

b $4p \times -3p$

4 Solve the following, leaving your answer as a fraction.

a $3x + 7 = 0$

b $4 - 5x = 0$

Incorporating exercises:	20A, 20B, 20C, 20D	**Key words**	
Homework:	20.1	coefficient	quadratic
Examples:	20.1	linear	expression

Learning objective(s)

● expand two linear brackets to obtain a quadratic expression

Prior knowledge

Pupils should be able to multiply out a single bracket.

Starter

Revise multiplying out a single bracket, for example $5(2x + 3)$, $x(2x + 3)$, $5x(2x + 3)$.

Main teaching points

Explain that a quadratic expression is one where the highest power of its terms is 2, for example x^2, $4x^2$, or $2x^2 + 3x$. When two brackets, each containing an x term, are multiplied together, the result is a quadratic expression. (Assuming there are no higher powers of x.)

If you multiply two brackets, each with two terms, there must be four multiplications.

In addition to the method of multiplying out the brackets shown in the text there are two other methods that are commonly used. Pupils who struggle with one method may have success with another (see Worked examples 12.1).

Common mistakes

Pupils often have problems with the negative signs when expanding.

Plenary

Complete an example using the different methods pupils have learned, or if you have shown one method, put a few questions on the board and ask for help from the pupils in completing the expansions.

Incorporating exercises:	20E, 20F, 20G	Key words	
Homework:	20.2	brackets	factorisation
Examples:	20.2	coefficient	quadratic
		difference of	expression
		two squares	

Learning objective(s)

● factorise a quadratic expression into two linear brackets

Prior knowledge

Pupils should be confident in factorising simple expressions, for example $5x + 10$, $5x^2 + 10x$, $5xy - 20x$.

Starter

Ask the pupils to factorise:
● $8r + 2t$
● $12s + 6t$
● $4a - 12ab$
● $16ab + 12b$
● $7acd - 5abc$
● $3t^2 + 6t$
● $12ab^2 + 24b$
● $9x^3 - 27xy$

Main teaching points

Recognising which method to use is often the hardest part for pupils. When they have completed exercises 20E, 20F, and 20G, go back and look at the questions with them. Pick out points that show them which method to use.

Common mistakes

Using two negatives in the difference of two squares. Making sure the 'x^2' part and the '+ number' part are correct, but not checking the 'x' part is correct.

Differentiation

Pupils must learn simple quadratic factorising and the difference of two squares method. You may decide that some pupils should by-pass more complicated factorisation (for example, $7x^2 - 37x + 10$) and go directly to Section 12.4.

Plenary

Write different quadratic expressions on the board and ask pupils which method they should use.

Incorporating exercises:	20H, 20I
Homework:	20.3
Example:	20.3

Key words
factors
solve

Learning objective(s)

● solve a quadratic equation by factorisation

Prior knowledge

Pupils need to be able to factorise the quadratic expressions met in exercises 20E, 20F and 20G.

Starter

Ask the pupils to factorise some of the questions (without the '= 0' part) from Exercise 20H (questions 13–27).

Main teaching points

Before factorising, ensure the equation ends '= 0', and is cancelled down as much as possible. Once factorised, write out the options, for example for $(x + 2)(x - 3) = 0$, write $x + 2 = 0$ or $x - 3 = 0$ before solving to get $x = -2$ or $x = 3$. Hopefully, this will stop pupils giving incorrect answers of $x = 2$ and $x = -3$.

Common mistakes

See main teaching point above.

Differentiation

For those who just cannot seem to factorise, tell them there are other ways to solve quadratics which will be covered later.

Plenary

Put $8x^2 - 3 = 2x^2 - 11x - 6$ on the board and get volunteers to give one step at a time to solve the quadratic. (Factorises to $(3x + 1)(2x + 3) = 0$.)

Incorporating exercise:	20J		Key words	
Homework:	20.4		quadratic	soluble
Example:	20.4		formula	solve

Learning objective(s)

● solving a quadratic equation by using the quadratic formula

Prior knowledge

Pupils need to be able to substitute into formulae, understand what the implications of the ± sign are, and have good calculator skills.

Starter

Pose the following questions:
● $5 \pm 10 = ?$ (15 or –5)

● $\dfrac{5 \pm 10}{3} = ?$ (5 or $-\dfrac{5}{3}$)

● $\dfrac{5 \pm \sqrt{10}}{3} = ?$ (2.72 or 0.61)

Main teaching points

There are a lot of things happening all at once here – many pupils will be tempted to go for the easy option and just give answers. Don't let them – there are typically four or five GCSE marks (grade A) for a question on solving using the formula. One mistake could result in zero if they only write answers. Demand that pupils show full step-by-step workings.

Common mistakes

Not quoting the formula before attempting to use it – one substitution error will then get no marks as the correct formula has not been shown.

When using the calculator, not pressing '=' before dividing by '2a'.

For example, for $x = \dfrac{4 + \sqrt{40}}{2}$ on the calculator:

$4 + \sqrt{40} \div 2 = (=7.16\ldots)$ rather than $4 + \sqrt{40} =$, then $\div 2 = (=5.16\ldots)$

or (even better) $(4 + \sqrt{40})$, then $\div 2 = (= 5.16\ldots)$

Pupils after ignore the negatives, so if solving $x^2 + 4x - 28 = 0$, using $a = 1$, $b = 4$, and $c = 28$ (not –28).

Differentiation

Given time and practice, even those that could not factorise will usually be able to complete this type of question as long as full workings are shown, and pupils do not take shortcuts.

Plenary

Choose a complicated question and go over the formula method, showing all steps and reinforcing the need to write down all steps to get as many marks as possible in the GCSE.

Incorporating exercise:	20K		**Key words**	
Homework:	20.5		completing	surd
Examples:	20.5		the square	
			square root	

Learning objective(s)

⬤ solve a quadratic equation by completing the square

Prior knowledge

Being able to recognise and factorise using the difference of two squares is very useful (Exercise 12F).

Starter

Ask the following questions:
⬤ Factorise $x^2 - 4$. $((x - 2)(x + 2))$
⬤ Expand $(x - 2)^2$. $(x^2 - 4x + 4)$
⬤ Expand and simplify
  a $(x + 2)^2 - 2^2$ $(x^2 + 4x)$
  b $(x + 3)^2 - 3^2$ $(x^2 + 6x)$
  c Guess $(x + 5)^2 - 5^2$ $(x^2 + 10x)$

Main teaching points

If you have to complete the square on $x^2 + ?x \ldots$, always halve the ? to get the number at the end of the bracket, that is, $(x + \frac{?}{2})^2$.

If you have to complete the square on $2x^2 + \ldots$, always factor out the 2 first, so $2(x^2 + \ldots)$ carrying on to get $2[(x + \frac{?}{2})^2 \ldots]$.

If you have to complete the square on $3x^2 + \ldots$, always factor out the 3 first, so $3(x^2 + \ldots)$ carrying on to get $3[(x + \frac{?}{2})^2 \ldots]$, etc.

When solving, always remember that $\sqrt{?} = \pm$ something.

Common mistakes

Not remembering that, for example, $\sqrt{9} = \pm 3$, and not leaving answers in surd form if asked to give an exact answer.

Plenary

Get pupils to help solve $x^2 + 4x = 0$, then $x^2 + 4x + 2 = 0$.
$(x = 0$ or -4, $x = \pm \sqrt{2} - 2$ $(= -0.59, -3.41))$

Incorporating exercises:	20L, 20M, 20N	**Key words**
Homework:	20.6	discriminant
Examples:	20.6	

Learning objective(s)

- recognise why some quadratic equations cannot be factorised
- solve practical problems using quadratic equations

Prior knowledge

Pupils should be able to use the quadratic formula, as well as the other ways of solving quadratic equations.

Starter

Give a variety of questions previously attempted in this chapter by the class as a quick(ish) revision. For example, Homework 20.3 question 4, Homework 20.5 question 1c, Homework 20.5 question 3 by formula.

Main teaching points

In the quadratic formula, '$b^2 - 4ac$' is called the discriminant.
If '$b^2 - 4ac$' is > 0, then there will be two solutions.
If '$b^2 - 4ac$' is $= 0$, then there will be only one solution.
If '$b^2 - 4ac$' is < 0, it cannot be solved at GCSE level.

Common mistakes

See those of 12.2 to 12.5.

Plenary

The perimeter of a rectangle is 34 cm, the diagonal is 13 cm. Find the width.

This (Pythagoras) question will test logical thinking and problem solving skills. The fairly easy numbers (5, 12, 13 triangle) are easy to cope with. Ask for step-by-step help from the pupils whilst solving the problem on the board.

Overview

21.1 Straight-line distance–time graphs
21.2 Straight-line velocity–time graphs
21.3 Other types of graphs

Sections 21.1 and 21.2 are an introduction to distance–time and velocity–time graphs, and how to interpret them. The last section includes a variety of other graphs including the 'filling a container' graph so popular with examiners.

Context

Distance–time graphs have some use in determining journey times but their main use, and that of velocity–time graphs, is in science and higher maths (mechanics). Section 21.3 contains graphs which help to calculate tax and mortgage payments – both unfortunately relevant to most of us.

AQA B references

AO2 Number and algebra: Sequences, functions and graphs

21.1, 21.2, 21.3   2.6d "…discuss and interpret graphs modelling real situations"

Route mapping

Exercise	D	C	B	A	A*
A	all				
B		all			
C			all		
D			1	2–3	

Answers to diagnostic Check-in test

1 a 2 m   **b** 3 hours   **c** 2 m/h

2 a Fills very fast at an even rate.

 b Fills at an even rate, fairly fast, but not as fast as **a**.

 c Fills slowly, at an even rate.

 d Fills slowly at first at an even rate, then very fast at an even rate.

 e Fills slowly initially, but the rate changes, getting faster and faster – not even.

1 This is a graph showing how far Cedric the slug goes across his lawn.

  a How far does Cedric go in 1 hour?

  b How long does it take Cedric to travel 6 metres?

  c What is Cedric's speed in metres per hour (m/h)?

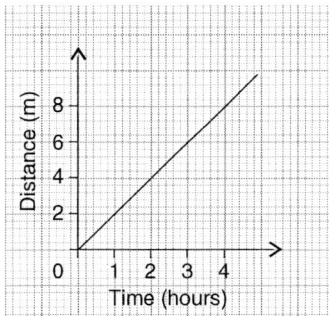

2 The following shapes are all containers that are to be filled with water from a hosepipe. The water flows at a steady rate all of the time.

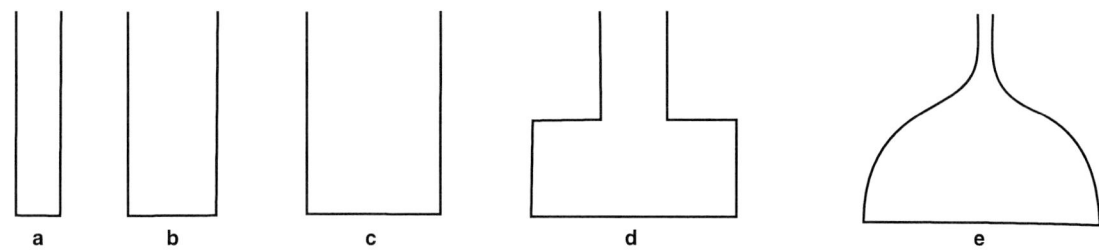

Describe what happens to the depth of the water in each container as it fills. Does the depth of the water increase quickly, slowly, at a constant rate, or at a changing rate?

Incorporating exercises:	21A, 21B	Key words	
Homework:	21.1	distance	speed
Example:	21.1	gradient	time

Learning objective(s)

● interpret distance–time graphs

Prior knowledge

Pupils need to be confident in plotting points on coordinate axes, as well as being able to read scales on axes. The majority of mistakes made on these questions involve misreading the scales.

Starter

Ask the class the following questions:
● If I walk 200 metres in 1 minute, what is my speed? Change from m/min, to m/h, to km/h.
● If I run 200 m in 20 seconds, what is my speed? Change from m/s, to m/min, to m/h, to km/h.
● If I fell from a plane, my speed might be 180 km/h. Change this to km/min, to km/s, to m/s.
● How would I sketch a graph showing a speed of 180 km/hr?

Main teaching points

Pupils should use the graph (or other information) to provide the units for example, m/s, km/h, etc. The steeper the graph, the faster the object. A flat part of the graph means the object is stationary.

Common mistakes

Incorrectly converting between units, for example making errors in converting from m/s to km/h. Also, misreading the scales used on the graphs.

Plenary

Choose any questions from the exercises completed by the pupils, and ask for the important points and methods required.

Incorporating exercise:	21C	**Key words**	
Homework:	21.2	acceleration	time
Examples:	21.2	gradient	velocity

Learning objective(s)

● interpret velocity–time graphs

Prior knowledge

Being practiced at using straight-line distance–time graphs will accelerate understanding in this section.

Starter

● What is happening on this distance–time graph?
  a Moving with uniform speed.
  b Stationary
  c Going back to the starting point at a uniform Speed which is slower than in part **a**.

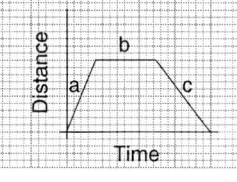

● What is happening on this velocity–time graph?
  a Getting faster and faster – i.e. accelerating.
  b Travelling at a constant speed.
  c Slowing down – i.e. decelerating, but with a slower rate of change than in part **a**.

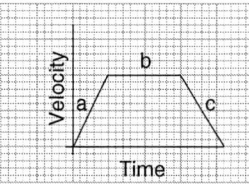

Main teaching points

Use the graph, or other information, to work out the units for acceleration (for example, m/s^2 or ms^{-2}, km/h^2 or kmh^{-2}). The steeper the graph, the faster the acceleration. A flat part to the graph, means constant velocity, that is, no acceleration.

Common mistakes

Misreading the scales, and incorrectly converting between units.

Plenary

Draw a set of axes on the board for velocity against time, but use no numbers. Using a car as a real-life example, ask the pupils to explain what is happening in each of these sections of graphs.

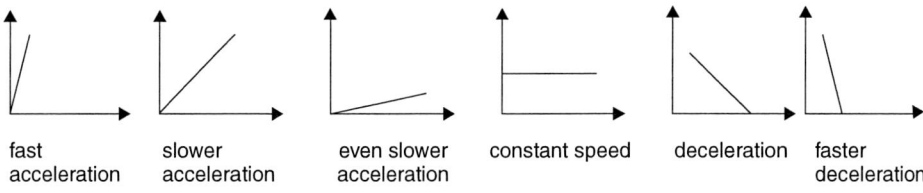

fast acceleration　　slower acceleration　　even slower acceleration　　constant speed　　deceleration　　faster deceleration

Incorporating exercise:	21D	
Homework:	21.3	
Examples:	21.3	

Learning objective(s)

● identify and draw some of the more unusual types of real-life graphs

Prior knowledge

A variety of maths skills (especially calculator skills) can be brought to bear here, as can the imagination when various vessels are filled with water.

Starter

With the following containers, ask the pupils to explain about the speed at which the height of the water increases, as they are filled with water from a tap that is flowing at a steady rate. Don't draw the graphs.

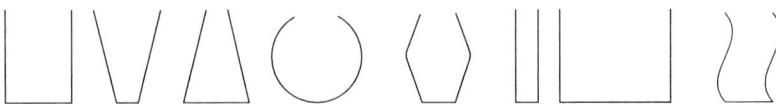

Main teaching points

Careful calculations need to be done on tax, mortgage and other money questions, as these will have to be plotted. With the 'filling the container' questions, pupils should remember that the flow in will be at the same rate and, the wider the container, the slower the rate of change in the height of the water.

Common mistakes

Poor calculator use. Misinterpretation of flow rates and, for example, giving the graph for _/ instead of /\‾\.

Plenary

Draw different shaped containers and get the pupils to draw graphs on the board. Alternatively, ask pupils to test each other by drawing their own container on the board.

Draw a graph and get pupils to work out the shape of the container.

This chapter covers all trigonometry
required at higher level, from
trigonometry of right-angled
triangles through to the sine and
cosine rule and their applications.

AQA A references

AO3 Shape, space and measures: Geometrical reasoning

22.1–22.2 3.2g "understand, recall and use Pythagoras' theorem in 2-D, then 3-D problems; investigate the geometry of cuboids including cubes, and shapes made from cuboids, including the use of Pythagoras' theorem to calculate lengths in three dimensions"

22.3–22.6 3.2h "... understand, recall and use trigonometrical relationships in right-angled triangles, and use these to solve problems, including those involving bearings, then use these relationships in 3-D contexts, including finding the angles between a line and a plane (but not the angle between two planes or between two skew lines); calculate the area of a triangle using $\frac{1}{2}ab\sin C$; draw, sketch and describe the graphs of trigonometric functions for angles of any size, including transformations involving scalings in either or both the x and y directions; use the sine and cosine rules to solve 2-D and 3-D problems"

Route mapping

Exercise	D	C	B	A	A*
A				1–4	5–6
B					all
C					all
D					all
E					all
F					all
G				1–2	3–11
H				1–5	6–11
I				1–2	3–5
J					all
K				1–3	4–9

Answers to diagnostic Check-in test

1 a 6.13 cm   **b** 2.89 cm   **c** 10.0 cm   **d** 10.4 cm   **e** 3.46 cm
   f 38.7°   **g** 30.5°   **h** 15.3 cm   **i** 22.4 cm

2 a 30 cm²   **b** 18.3 cm²   **c** 39.9 cm²   **d** 38.2 cm²

1 Find the angles or sides marked *x* in these right-angled triangles.

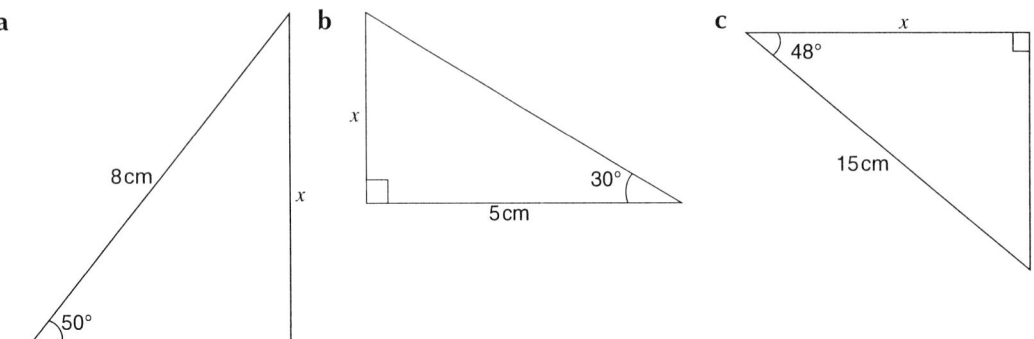

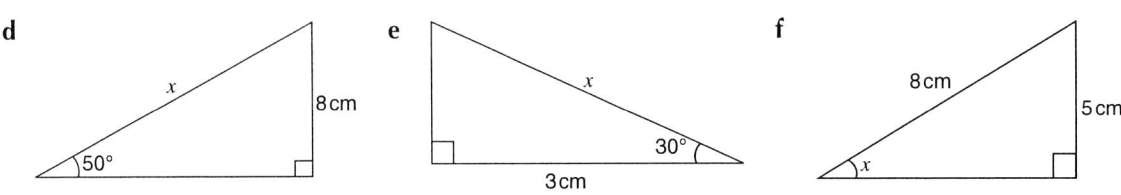

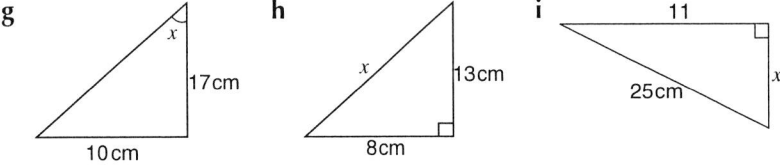

Module 5: Algebra and Space, shape and measure

2 Calculate the areas of these triangles.

a

5cm

12cm

b

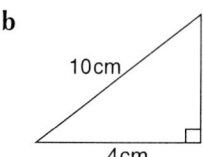

10cm

4cm

c

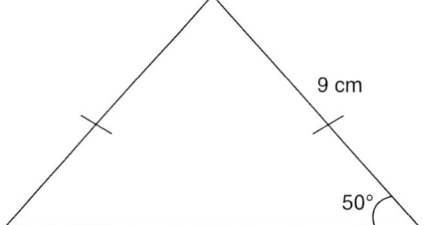

9 cm

50°

d

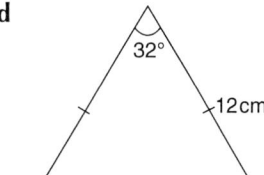

32°

12cm

Incorporating exercise:	22A	**Key words**	
Homework:	22.1	area	
Examples:	22.1	length	
		perpendicular	

Learning objective(s)

● use trigonometric ratios and Pythagoras' theorem to solve more complex two-dimensional problems

Prior knowledge

Pupils should know how to find the sides of right-angled triangles using Pythagoras' theorem. They should also be able to find the missing angles and sides of right-angled triangles using sine, cosine and tangent.

Starter

Ensure pupils successfully complete the check-in test, particularly question 1.

Remind pupils of the properties of certain polygons through the use of class questioning.

What do the interior angles of a pentagon add up to?

What can you tell me about the angles and sides in a rhombus?

What is special about the angles and sides in an isosceles triangle?

Main teaching points

There is no new mathematical content in this section, which relies upon the use of sin, cos and tan in right-angled triangles and Pythagoras' theorem as covered in Section 12.3.

As always in working through trigonometric problems involving Pythagoras, a clear diagram is essential. Pupils often need to be reminded that full working out should be shown in each stage of the question.

Plenary

The section is designed to consolidate pupils' knowledge. It is suggested that the homework examples are gone through in class to ensure as many pupils as possible have an understanding of the different types of approach needed to successfully solve basic trigonometric problems.

Incorporating exercise:	22B
Homework:	22.2
Examples:	22.2

Learning objective(s)

● use trigonometric ratios and Pythagoras' theorem to solve more complex three-dimensional problems

Prior knowledge

Pupils should be confident in applying trigonometric ratios and Pythagoras' theorem to solve problems in two dimensions, as covered in Section 22.1.

Starter

It is suggested that the idea of the angle between a line and a plane is demonstrated to the pupils in the class. Ask pupils to try to define the angle between a line and a plane. The crucial thing for them to introduce themselves will be a line that is perpendicular to the plane in their definition. Only the most able are likely to formalise their ideas in this way.

Main teaching points

For each question, or part of question, in this section's Pupil Book exercises, it is usually necessary to extract and redraw a right-angled triangle from the diagram to enable a missing length or angle to be found. Pupils should get in the habit of doing this from the first question.

The angle between a line and a plane is often required to be found. Pupils can be confused as to what this angle actually is. Consider a general plane and a line going through it (see diagram below). Then the angle required is $\angle PXY$, where P lies on the plane such that $\angle XPY = 90°$.

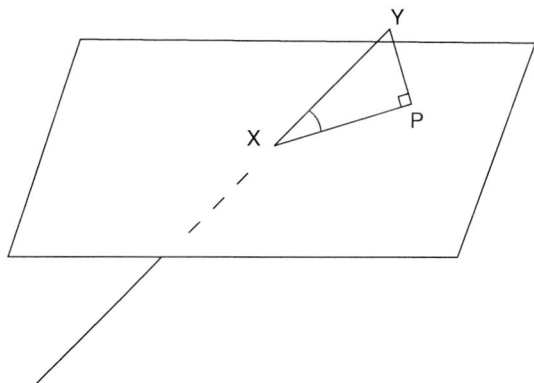

Common mistakes

The most common mistakes are to either pick out the wrong triangle required, or not pick out anything at all. Care must be taken to label all known quantities in the new triangle and to label the unknown side or angle with a letter.

Plenary

For a plenary, it is again suggested that the homework examples are gone through in class. Question 4 of Exercise 22B in the Pupil Book in particular focuses directly on the angle between a line and a plane, and it is with the skill of picking out the angle required where the most difficulties may lie.

Incorporating exercises:	22C, 22D, 22E, 22F	Key words
Homework:	22.3	cosine
Examples:	22.3	sine
		tangent

Learning objective(s)

● find the sine, cosine and tangent of any angle from 0° to 360°

Prior knowledge

Pupils should be confident in applying sine, cosine and tangent to problems in right-angled triangles, as covered in Sections 22.1–22.2.

Starter

The discovery activities in the Pupil Book (on pages 475 and 477) are ideal starter activities. It is suggested that pupils initially produce accurately plotted graphs for $y = \sin x$, $y = \cos x$ and $y = \tan x$ for $0° \leq x \leq 360°$, each filling, as far as possible, a side of A4 graph paper in landscape mode. These graphs can then be used as a reference for work in this section.

When plotting $y = \tan x$ pupils will inevitably ask why tan 90° and tan 270° give error messages on their calculator. Suggest they leave these values out and plot the graph first. The idea of an asymptote could then be discussed later.

Main teaching points

Impress upon the pupils the need to learn the shapes and characteristics of the graphs $y = \sin x$ and $y = \cos x$ for $0° \leq x \leq 360°$, as well as their respective ranges. Note that the graph of $y = \tan x$ is no longer required at GCSE, but a discussion of its characteristics, including asymptotes, would be a useful inclusion alongside the sine and cosine curves.

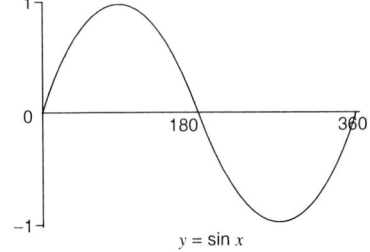
$y = \sin x$

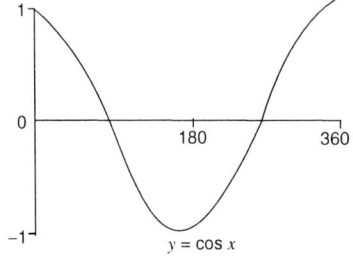
$y = \cos x$

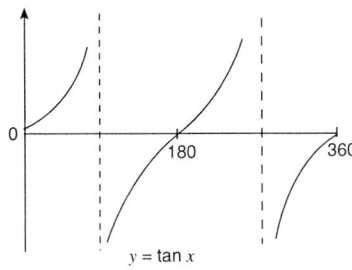
$y = \tan x$

Emphasise the difference between sketch graphs (those shown here) and accurately plotted graphs (those that they produced). Being able to produce the sketch graphs themselves is not only a requirement in the higher syllabus, it will also be one of the most useful ways of solving simple trigonometric equations through the apparent immediacy of the symmetries of the graphs.

The activities suggested lead the pupils towards formalising the trigonometric ratios for any angle in terms of the ratio of the respective acute angle. So, for:

● $90° < x < 180°$, $\sin x = \sin(180 - x)$, $\cos x = -\cos(180 - x)$, $\tan x = -\tan(180 - x)$
● $180° < x < 270°$, $\sin x = -\sin(x - 180)$, $\cos x = -\cos(x - 180)$, $\tan x = \tan(x - 180)$
● $270° < x < 360°$, $\sin x = -\sin(360 - x)$, $\cos x = \cos(360 - x)$, $\tan x = -\tan(360 - x)$

It is unlikely that pupils will need or be able to memorise these. However, providing one or two specific examples (such as sin 30 = sin 150) will demonstrate their use as a point of reference.

Plenary

Ask pupils for suggestions as to what happens to trigonometric ratios of angles larger than 360°.
By drawing a sketch graph, ask them to solve the equation:

$$\cos x = \frac{1}{2} \text{ for } 0° \leq x \leq 720°$$

Can they predict any further solutions?

If we know one solution, for instance $x = 60°$ in this case, can they suggest a general formula for the solution of:

$$\cos x = \frac{1}{2}$$

Try and guide them towards the expression $360n \pm 60$, perhaps verbally at first.

			Key words
Incorporating exercises:	22G, 22H, 22I		cosine rule
Homework:	22.4a, 22.4b		included angle
Examples:	22.4		sine rule

Learning objective(s)

● find the sides and angles of any triangle whether it has a right angle or not

Prior knowledge

Pupils should know how to calculate the angles and sides of right-angled triangles using sine, cosine and tangent and Pythagoras' theorem.

Starter

Illustrate what is in effect the sine rule, from a specific example. Draw the triangle on the board. Ask pupils for ideas on how they might find the length of side b.

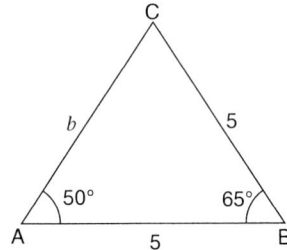

Start by asking why we cannot use sin, cos or tan in this case. (They can only be used in right-angled triangles.) Then suggest pupils consider what happens if we 'make' a right-angled triangle by dropping a perpendicular from vertex C. Now can pupils see how they might find the length of line AC?

Eventually this should lead to the appreciation that $b \sin 50 = 5 \sin 65$ and, therefore, that $\dfrac{b}{\sin 65} = \dfrac{5}{\sin 50}$.

Point out that the equal ratios are the lengths of the sides multiplied by the sines of the opposite angles. Is this always true? Investigate by forming a general triangle and thus deriving the sine rule.

Main teaching points

Every non right-angled triangle carries six pieces of information, namely three angles and three sides. Given any three pieces of information (for example, two angles and one side) the aim is to determine possibilities for the remaining three (using the sine and cosine rules).

A crucial introduction to this section is the idea of labelling the triangle. Following on from the work done on right-angled triangles, it is important to stress the importance of labelling the vertices of a triangle with capital letters and those of the opposite sides with the corresponding letters in lower case.

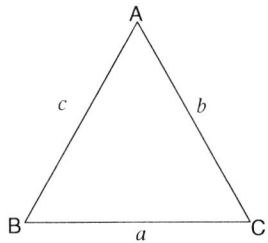

To avoid potential pitfalls with algebraic manipulation, it should be emphasised that, when working with the sine rule:

- use $\dfrac{a}{\sin A} = \dfrac{b}{\sin B} = \dfrac{c}{\sin C}$ when calculating a missing side;

- use $\dfrac{\sin A}{a} = \dfrac{\sin B}{b} = \dfrac{\sin C}{c}$ when calculating a missing angle.

Once pupils are comfortable with the sine rule, and have tried some questions, guide them carefully through the cosine rule.

Consider the following triangle:

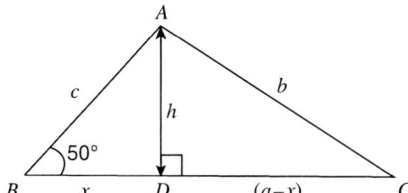

By Pythagoras' theorem we have $h^2 = c^2 - x^2$ and $h^2 = b^2 - (a - x)^2$.

Solving simultaneously we obtain $c^2 - x^2 = b^2 - (a - x)^2 = b^2 - a^2 + 2ax - x^2$.
Therefore $c^2 = b^2 - a^2 + 2ax$.

But $x = c \cos B$ and so $c^2 = b^2 - a^2 + 2ac \cos B$, or $b^2 = a^2 + c^2 - 2ac \cos B$.

From the symmetry of the triangle we can therefore deduce the three standard forms of the cosine rule:

$a^2 = b^2 + c^2 - 2bc \cos A$
$b^2 = a^2 + c^2 - 2ac \cos B$
$c^2 = a^2 + b^2 - 2ab \cos C$

Again, pupils should always be encouraged to correctly label the triangle and write down the general and appropriate form of the cosine rule for every question attempted. As well as being good mathematical practice, this also aids the pupil's ability to memorise the cosine rule in its various forms.

More able pupils could be encouraged to spot the pattern of the cosine rule so that the required calculation might be written down straight away, as done in the worked examples. This approach however is only to be recommended once the cosine rule is thoroughly understood.

Common mistakes

Pupils' main difficulty with the cosine rule is more often than not their obtaining the incorrect answer through the misuse of their calculator. This can occur especially when the angle whose cosine is being taken is obtuse, and therefore the cosine is negative. It can help if the formula is written in this form (that is, with brackets around the final term).

$a^2 = b^2 + c^2 - (2bc \cos A)$

Encourage less able pupils to evaluate the bracket first, including extra lines in their working if necessary as shown in Worked example 22.4b.

Plenary

Solving the triangle

In general, it is easier for pupils to use the sine rule rather than the cosine rule, and it is also true that the cosine rule does not need to be used more than once in the process of solving the triangle. It is, however, useful to discuss the advantages and disadvantages of using the cosine rule.

Suppose, for example, you have a triangle with sides 8 cm, 12 cm and 15 cm.

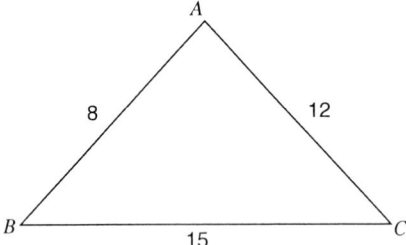

In completing the triangle, you first need to apply the cosine rule. Suppose you calculate

$$\cos A = \frac{8^2 + 12^2 - 15^2}{2 \times 8 \times 12} = -\frac{17}{192}.$$ Therefore $A = 95.1°$.

For the second step, it may well be easier to use angle A in an application of the sine rule to find one of the remaining angles. However, more able pupils should be aware that this might incur a loss of accuracy (through using 95.1°) which can be avoided if we go back and apply the cosine rule to the information originally given in the question.

In most cases, of course, this will make little or no difference to final answers, though in a higher level set of a relative range of ability, answers from pupils are often obtained that differ in the third significant figure, and obtaining a collection of such answers may be a useful introduction to discussing the idea of loss of accuracy.

The complete sine rule

Should time permit, it is worth a look at the complete sine rule. Although this result is not required at GCSE, it is given here for sake of completeness. The proof in particular is accessible to A/A* pupils.

The circumcircle is a triangle's circumscribed circle, the unique circle that passes through each of the triangle's three vertices. The centre O of the circumcircle is called the circumcentre, and the circle's radius r is called the circumradius. The complete sine rule is given by:

$$\frac{a}{\sin A} = \frac{b}{\sin B} = \frac{c}{\sin C} = 2r$$

It can be proved as follows. Let O be the centre of the circumcircle of triangle ABC and let M be the midpoint of BC.

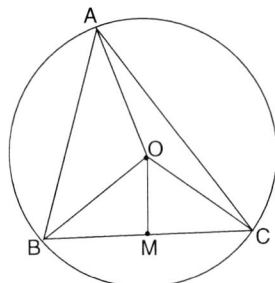

Then $\angle MOC = \frac{1}{2}\angle BOC = \frac{1}{2}(2\angle BAC) = \angle BAC$.

Therefore, $\sin A = \frac{MC}{OC} = \frac{\frac{a}{2}}{r}$ and so $\frac{a}{\sin A} = 2r$. The complete sine rule then follows.

22.5   Sine, cosine and tangent of 30°, 45° and 60°

Incorporating exercise:	22J
Homework:	22.5
Examples:	22.5

Learning objective(s)

● work out the trigonometric ratios of 30°, 45° and 60° in surd form

Prior knowledge

Pupils should be familiar with the use of the trigonometric ratios sine, cosine and tangent in relation to right-angled triangles.

Starter

Examples 17 and 18 in the Pupil Book clearly demonstrate the method of finding the trigonometric ratios of 30°, 45° and 60° in surd form, using an equilateral triangle and an isosceles triangle. These examples should be gone through with the class.

Main teaching points

The exact trigonometric values obtained will rarely be used in any practical application at GCSE. Rather, the emphasis is on demonstrating how trigonometric ratios of angles may be found without first finding the angles themselves. This will most likely involve a single application of Pythagoras' theorem as shown by those questions comprising Exercise 22J in the Pupil Book.

Plenary

Ask the class to find the exact value for the area of an equilateral triangle of length a cm. Given that $\sin x = k$, find exact values for $\cos k$ and $\tan k$.

Incorporating exercise:	22K
Homework:	22.6
Example:	22.6

Learning objective(s)

● calculate the area of a triangle given two sides and the included angle

Prior knowledge

Pupils should have covered the sine and cosine rules in Section 22.4.

Starter

Illustrate a specific example of the formula for finding the area of a triangle using two sides and the included angle by working from first principles. Show pupils, for example, this triangle:

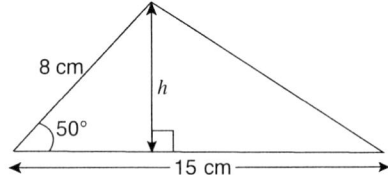

The area of the triangle is given by $A = \frac{1}{2} \times 15 \times h$, but we need to know h. This can be found from $\sin 50 = \dfrac{h}{8}$.

So, $A = \frac{1}{2} \times 15 \times 8 \times \sin 50 = 46.0 \, \text{cm}^2$.

Main teaching points

The key point, obviously, is for pupils to know and be able to use the formula:

$$A = \frac{1}{2} ab \sin C$$

One way for pupils to remember this is through the form: area equals half the product of any two sides times the sine of the *included* angle.

If two sides and a non-included angle are given, the included angle may be found from a single application of the sine rule (which would find the other non-included angle), followed by subtracting the size of these two non-included angles from 180°.

For a triangle with three given sides, a missing included angle may be found through a single application of the cosine rule.

Plenary

A suggested plenary is a demonstration of Hero's formula, which allows the area of any triangle to be found if the three sides are known. Hero's formula states the area of a triangle is:

$$A = \sqrt{s(s-a)(s-b)(s-c)} \text{ where } s = \frac{a+b+c}{2}$$

For a triangle with three given sides, it is straightforward enough for higher tier candidates to find the area through an application of the cosine rule (to find a missing angle) followed by the use of the sine formula for finding the area using two sides and an included angle.

Hero's formula, while not explicitly part of the higher GCSE syllabus is an entirely valid and attractive alternative method that many candidates might look to use. This proof is not often encountered and yet should be accessible for A* students at GCSE. Start, by considering this triangle:

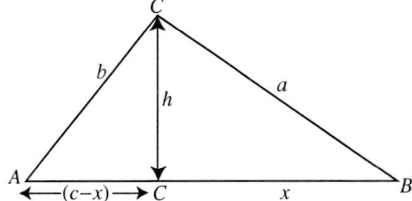

Let $s = \dfrac{a+b+c}{2}$. From Pythagoras' theorem, we have:

(1) $x^2 + h^2 = a^2$

(2) $(c - x)^2 + h^2 = b^2$

Solving these simultaneously gives:

$x^2 - (c - x)^2 = a^2 - b^2$, or $2cx = a^2 - b^2 + c^2$

From equation 1:

$$4c^2h^2 = 4a^2c^2 - 4c^2x^2$$

$$= (2ac + 2cx)(2ac - 2cx)$$

$$= (2ac + a^2 - b^2 + c^2)(2ac - a^2 + b^2 - c^2)$$

$$= ((a + c)^2 - b^2)((b^2 - (a - c)^2)$$

$$= (a + c + b)(a + c - b)(b + a - c)(b - a + c)$$

$$= 2s(2s - 2b)(2s - 2c)(2s - 2a)$$

$$= 16s(s - a)(s - b)(s - c)$$

Now to find the area of the triangle:

$$A = \frac{1}{2}ch = \sqrt{\frac{4c^2h^2}{16}} = \sqrt{s(s-a)(s-b)(s-c)}$$

Overview

23.1 Linear graphs **23.2** Finding the equation of a line from its graph **23.3** Uses of graphs **23.4** Parallel and perpendicular lines	This chapter covers all the methods (grades D, C, and B) for drawing linear graphs, finding their equations, and a few uses for them.

Context

Much of our knowledge and use of science is, or can be, displayed in graph form – everything from braking distances of cars, to break-even points of a company's economic future.

AQA B references

AO2 Number and algebra: Equations, formulae and identities

23.3	2.5j "… interpret [linear simultaneous] equations as lines and their common solution as the point of intersection"

AO2 Number and algebra: Sequences, functions and graphs

23.1, 23.2, 23.3, 23.4	2.6b "use conventions for coordinates in the plane; plot points in all four quadrants; recognise … that equations of the form $y = mx + c$ correspond to straight-line graphs in the coordinate plane; plot graphs of functions in which y is given explicitly in terms of x, or implicitly; …"
23.2, 23.4	2.6c "find the gradient of lines given by equations of the form $y = mx + c$ …; understand that the form $y = mx + c$ represents a straight line and that m is the gradient of the line and c is the value of the y-intercept; explore the gradients of parallel lines and lines perpendicular to each other"
23.3	2.6d "construct linear functions … arising from real-life problems;…"

Route mapping

Exercise	D	C	B	A	A*
A	all				
B		all			
C		all			
D			all		
E			all		
F			all		
G			all		
H				1–4	5–10

1 A = (1, 1) B = (−1, 3) C = (−3, −1) D = (1, −2) **2** $y = 4$ **3** $y = -5$

4

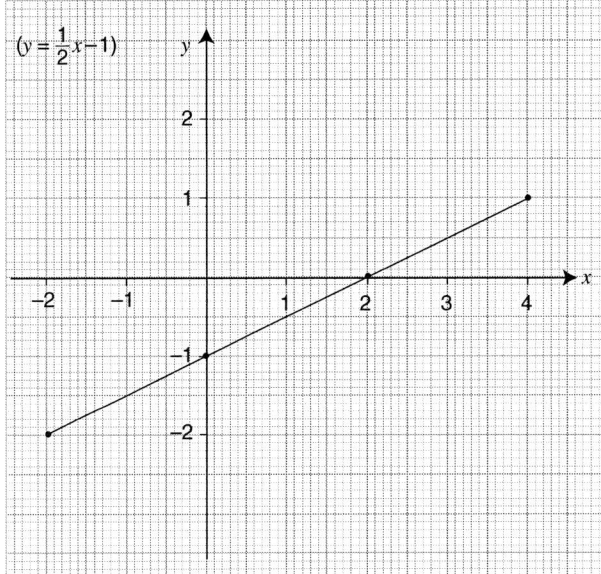

5

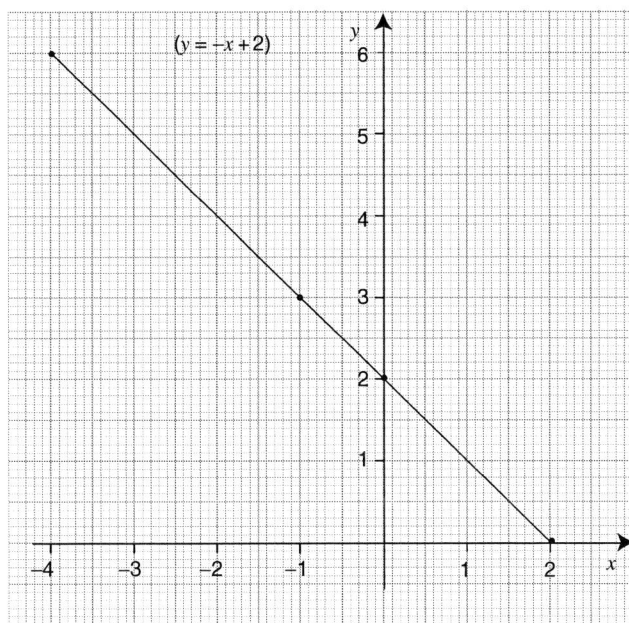

1 What are the coordinates of the points shown on the grid?

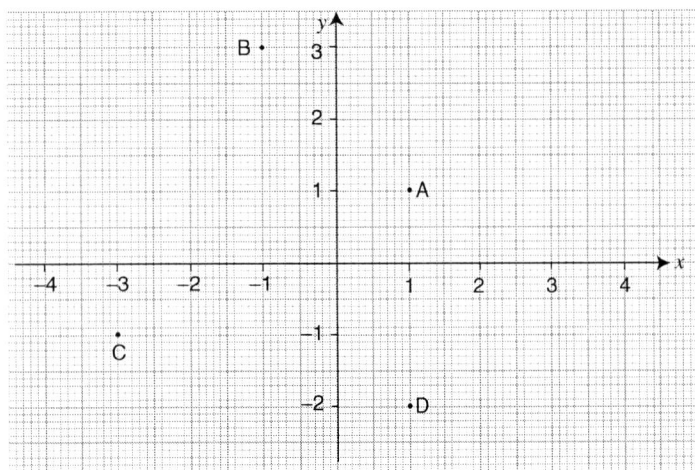

A =
B =
C =
D =

2 Find the value of y when $x = 3$, using the rule $y = 2x - 2$.

3 Find the value of y when $x = -3$, using the rule $y = 2x + 1$.

4 Plot the points in the table on the grid below. Join the points with a straight line.

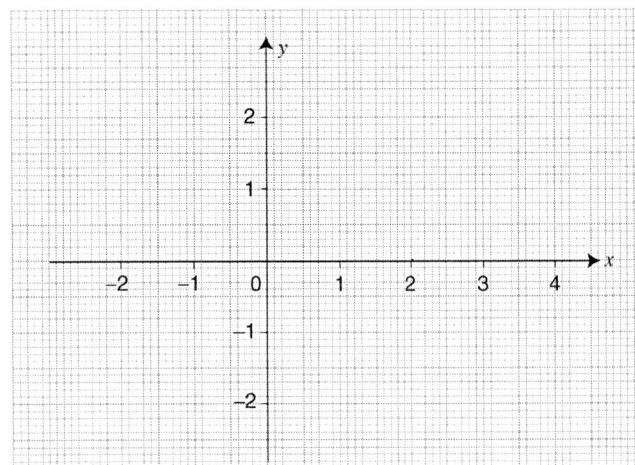

x	-2	0	2	4
y	-2	-1	0	1

5 Plot the points $(-4, 6)$, $(-1, 3)$, $(0, 2)$, $(2, 0)$ on the grid below. Join the points with a straight line.

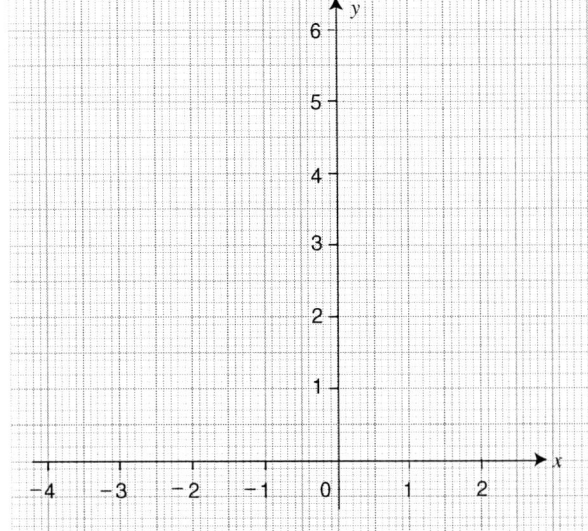

    Module 5: Algebra and Space, shape and measure

Incorporating exercises:	23A, 23B, 23C, 23D	Key words	
Homework:	23.1	axis (pl: axes)	linear graphs
Examples:	23.1	coefficient	
		gradient-intercept	

Learning objective(s)

● draw linear graphs without using flow diagrams

Prior knowledge

Pupils must be able to read and plot coordinates (even if given in table form). They must also be able to substitute into simple algebraic functions (Chapter 19, Section 19.1a).

Starter

Write a rule on the board, for example $\times 2 + 1$, and ask for the results when this rule is applied to 0, 1, 2, 3 and 4.
Give other rules and find the results, for example $\times 4 - 5$, $\div 2 - 3$.
Give other input numbers for these rules, for example -3, -2, -1.

Main teaching points

Pupils need to be able to input values given as both '-3 to $+3$' and '$-3 \leq x \leq 3$'. After drawing a graph, always write its equation on one end of it. When an equation is arranged as $y =$ something, for example, $y = 3x + 2$, the 3 is the steepness (gradient) of the line (1 unit across, 3 up), the $+2$ means the line will cut across the y-axis at $+2$.

Common mistakes

Working out two points and drawing the line — not checking with a third point that the line is correct. Algebra skills not being adequate.

Plenary

Draw a set of axes on the board.
Get a volunteer to draw the line with equation $y = 2x + 1$ (for example).
If it looks correct, ask for the (0, ?) and (?, 0) coordinates (and other points if suitable).
Repeat with other volunteers and other lines, for example $y = 3x - 2$, $y = 6x$, $y = \frac{1}{2} x + 4$.

Go over the different methods of drawing linear graphs — plotting points, gradient-intercept, and 'cover-up'.

			Key words	
Incorporating exercise:	23E		coefficient	intercept
Homework:	23.2		gradient	
Example:	23.2			

Learning objective(s)

- find the equation of a line using its gradient and intercept

Prior knowledge

Pupils should have understood and succeeded in Exercise 23C.

Starter

Draw a set of axes numbered from -5 to $+5$ on both axes.
- Draw the line $y = 2x + 1$ by plotting three points.
- Draw the line $y = 2x - 1$ by using the gradient-intercept method.
- What would you do if asked to draw $4x + 5y = 20$? (Use cover-up method and draw.)

Main teaching points

Tell pupils that for $y = mx + c$, the 'm' is the gradient. For example, in $y = 3x + 2$, $m = 3$, so the gradient $= 3$; so for every 1 unit the line goes to the right, it goes 3 units up. The 'c' is $+2$, so the intercept is 2; so the graph cuts the y-axis at $+2$.

Common mistakes

Missing the negative on the gradient, when the line is a negative gradient line.

Differentiation

Lower ability pupils find negative and fractional gradients much more of a challenge.

Plenary

Draw three or four sets of axes on the board and sketch a graph on each one. For example:
$y = 2x + 2$, $y = 2x - 4$, $y = -x + 2$, $y = \frac{1}{2}x$
DO NOT tell pupils the equations of the lines.
Ask the pupils to *explain* how to work out the equation of each line.

Incorporating exercises:	23F, 23G	Key words
Homework:	23.3	formula rule
Examples:	23.3	(pl: formulae)

Learning objective(s)

- use straight-line graphs to find formulae
- solve simultaneous linear equations using graphs

Prior knowledge

Pupils need to be able to give the equation of a line using gradient and intercept or cover-up method, and be able to draw a line given its equation.

Starter

Draw a line (for example $y = 2x - 3$) on the board, and ask the pupils to find the equation.
Ask the pupils to draw a set of axes (numbered 0 to 10 on both), and get them to draw the lines $y = 2x + 1$ and $x + y = 10$ on the same axes.
Ask the pupils to give the coordinates of the point of intersection of the two lines (answer should be (3, 7)). Use this to explain about solving simultaneous equations.

Main teaching points

Whenever possible, on conversion graph questions, draw the horizontal and vertical 'tracking' lines on the graph for maximum accuracy. For solving simultaneous equations, drawing graphs accurately is the main concern.

Common mistakes

When answering conversion graph questions, not drawing the horizontal and vertical lines often means accuracy is lost – using a finger is not good enough. When solving simultaneous equations questions, not taking time to ensure that the individual lines are drawn accurately.

Plenary

Draw this diagram on the board and show that $40x = 12y$:

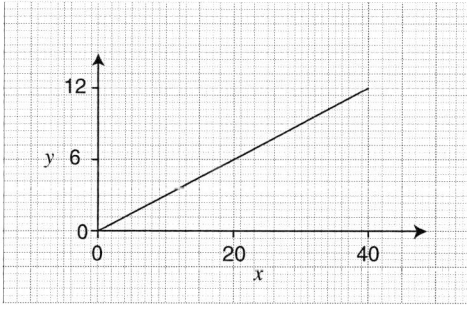

Ask, "What is y when $x = 100$?" (30)
"What is x when $y = 1200$?" (4000)

Incorporating exercise:	23H	**Key words**	
Homework:	23.4	negative	parallel
Examples:	23.4	reciprocal	perpendicular

Learning objective(s)

● draw linear graphs parallel or perpendicular to other lines and passing through a specific point

Prior knowledge

Pupils need to have been taught what a negative reciprocal is, and have had some practice at finding them. They must be familiar with gradient and intercept in $y = mx + c$, and they need to know what parallel and perpendicular lines are.

Starter

Ask the following questions:
● What does parallel mean?
● What does perpendicular mean?
● What do the m and c mean in $y = mx + c$?
● What is a reciprocal? What is the reciprocal of 2? ($\frac{1}{2}$)
● What is a negative reciprocal? What is the negative reciprocal of 3, -2, $\frac{1}{4}$, $-\frac{1}{2}$? ($-\frac{1}{3}$, $\frac{1}{2}$, -4, 2)

Main teaching points

Parallel lines have the same gradient (m) but different intercepts (c). Perpendicular lines have gradients which are the negative reciprocals of each other.

Common mistakes

Forgetting the negative when finding the gradient of a perpendicular line, that is, only finding the reciprocal.

Differentiation

The difference between A and A* questions is usually that A grade lines go through a given intercept.

Plenary

Write this equation on the board:
$y = 2x + 3$
Ask the pupils:
a What is the gradient of this line?
b What is the gradient of a parallel line?
c What is the gradient of a perpendicular line?

Ask the same questions for other lines such as: $y = -2x + 3$, $y = \frac{1}{2}x + 3$, $y = -\frac{1}{2}x - 1$

Other graphs

Overview

This chapter covers drawing and reading values from a variety of graphs.

Context

As with linear graphs, these more complex graphs have a huge variety of uses in science, economics, architecture, etc.

AQA B references

AO2 Number and algebra: Sequences, functions and graphs

24.1–24.4 2.6f "plot graphs of simple cubic functions, the reciprocal function $y = \frac{1}{x}$ with $x \neq 0$, the exponential function $y = k^x$ for integer values of x and simple positive values of k, the circular functions $y = \sin x$ and $y = \cos x$, ...; recognise the characteristic shapes of all these functions"

Route mapping

Exercise	D	C	B	A	A*
A				all	
B			1–5	6–8	
C				1–2	3–4
D					all

Answers to diagnostic Check-in test

1 7   **2** 40   **3** 8   **4** 0.866

1 For the equation $y = x^2 - 2$, find y when $x = 3$.

2 For the equation $y = x^3 + x^2 + x + 1$, find y when $x = 3$.

3 For the equation $y = 2^x$, find y when $x = 3$.

4 For the equation $y = \sin x$, find y when $x = 60°$.

   Module 5: Algebra and Space, shape and measure

Incorporating exerciss:	24A	Key words
Homework:	24.1	asymptote
Examples:	24.1	reciprocal
		square root

Learning objective(s)

● recognise square-root reciprocal and graphs

Prior knowledge

Good calculator skills are needed to complete the table, especially with some of the graphs. Drawing these graphs requires the same skills as for drawing quadratic graphs – one smooth curve passing through all points plotted.

Starter

Get the pupils to complete a table of values and draw a graph of a quadratic equation.

Main teaching points

Remind the pupils that the questions for graphs of square roots reciprocals and use the same methods as for quadratics – complete the table, plot the points, draw the curve. These graphs also tend to require more calculator use when completing the table.

Common mistakes

Poor calculator use – especially when calculating with negatives.

Plenary

Get volunteers to say or draw:
● the shape of $y^2 = x$, $y^2 = 4x$, $y^2 = 25x$, $y^2 = 100x$ (Sketch them on the same axes.)

● the shape of $y = \frac{1}{x}$, $y = \frac{2}{x}$, $y = \frac{3}{x}$ (Sketch them on the same axes.)

● what happens to each of the above equations if you put '+ 4' on the end

Incorporating exercis:	24B	**Key words**	
Homework:	24.2	cubic	
Examples:	24.2		

Learning objective(s)

● draw and recognise cubic graphs

Prior knowledge

Good calculator skills are needed to complete the table, especially with some of the graphs. Drawing these graphs requires the same skills as for drawing quadratic, square root and reciprocal graphs – one smooth curve passing through all points plotted.

Starter

Get the pupils to complete a table of values and draw a graph of a quadratic equation.

Main teaching points

Remind the pupils that the questions for graphs of cubics use the same methods as for quadratics – complete the table, plot the points, draw the curve. These graphs also tend to require more calculator use when completing the table.

Common mistakes

Poor calculator use – especially when calculating with negatives.

Plenary

Get volunteers to say or draw:

● the shape of $y = x^3$, $y = 2x^3$, $y = 3x^3$ (Sketch them on the same axes.)

What happens to each of the above equations if you put '+ 4' on the end?

Incorporating exercise:	24C	Key words
Homework:	24.3	exponential functions
Examples:	24.3	

Learning objective(s)

● draw and recognise exponential functions

Prior knowledge

Pupils must be able to use the power button on their calculators.

Starter

Give the pupils questions involving exponential growth and see if they can find the rule.
For example, bacteria double in number every 20 minutes. If you start with just one bacterium, how many would there be after:
a 1 hour?   **b** 6 hours?   **c** 12 hours?   **d** x hours? (**a** 8   **b** 262 144   **c** 6.87×10^{10} **d** 2^{3x})

Main teaching points

When calculating exponential growth (or decay), the numbers often get very big (or very small), very quickly.

Common mistakes

Incorrect use of the calculator when calculating values.

Differentiation

Some pupils will find 2^x relatively easy, but 2^{x-1} quite difficult to understand, so extra help may be needed here.

Plenary

What shapes are the graphs of $y = 2^x$, $y = 3^x$, $y = 4^x$, $y = (\frac{1}{2})^x$? (Sketch on the same axes.)

Incorporating exercise:	24D		**Key words**	
Homework:	24.4		cosine	
Examples:	24.4		sine	

Learning objective(s)

⬤ draw and recognise sine and cosine graphs

Prior knowledge

Pupils should have completed Chapter 22. They may still require reminders on calculator usage.

Starter

Ask a volunteer to draw the graph of $y = \sin x$ on the board – they may need some help initially. When completed, get all the pupils to copy the graph (as accurately as possible) into their books.
Repeat for $y = \cos x$.

Main teaching points

For the sine curve, show the symmetry of the first two quadrants (0° to 180°) about 90°, and the second two quadrants (180° to 360°) about 270°.

For the cosine curve, show the symmetry of the first and fourth quadrants, and the second and third quadrants about 180°.

Encourage the pupils to draw in pencil on the curves, when answering questions, to help their understanding of these symmetries.

Common mistakes

Drawing a wrong or incorrect graph; mixing up the sine and cosine graphs.

Plenary

⬤ Get a "volunteer" to sketch a sine curve on the board from 0° to 360°.
⬤ Ask another "volunteer" to show how the symmetries around 90° and 270° can be drawn on the curve to help in finding the second solution, if you are told that $\sin 70° = 0.939$ and $\sin 220° = -0.643$.

Overview

This chapter shows how to manipulate algebraic fractions. Section 25.2 deals with the solution of linear and non-linear simultaneous equations, Sections 25.3 to 25.5 show how to continue a number sequence, then how to find and use the nth term of linear and quadratic sequences. Finally, Section 25.6 brings together many algebraic manipulation skills in 'changing the subject'.

Context

In many ways, manipulation of algebraic expressions and changing the subject of a formula are the most important skills that a good mathematician can have, as it applies to so many areas of maths, as well as almost all of physics, engineering, economics, etc.

AQA B references

AO2 Number and algebra: Equations, formulae and identities

25.1, 25.2, 25.6	2.5a "distinguish the different roles played by letter symbols in algebra, using the correct notational conventions for multiplying or dividing by a given number, and knowing that letter symbols represent definite unknown numbers in equations, defined quantities or variables in formula, general, unspecified and independent numbers in identities, ..."
25.1, 25.2, 25.6	2.5b "... manipulate algebraic expressions by collecting like terms, multiplying a single term over a bracket, taking out common factors ..."
25.6	2.5h "... change the subject of a formula, including cases where the subject occurs twice, or where a power of the subject appears; ..."
25.2	2.5l "solve exactly, by elimination of an unknown two simultaneous equations in two unknowns, one of which is linear in each unknown, and the other is linear in one unknown and quadratic in the other or where the second is of the form $x^2 + y^2 = r^2$."

AO2 Number and algebra: Sequences, functions and graphs

25.3, 25.4	2.6a "generate common integer sequences ...; generate terms of a sequence using term-to-term and position-to-term definitions of the sequence; use linear expressions to describe the nth term of an arithmetic sequence ..."

Route mapping

Exercise	D	C	B	A	A*
A			1–4	5–7	8–9
B			1		2–5
C	1–5	6			
D		all			
E		all			
F			all		
G				1–13	14–16

Answers to diagnostic Check-in test

1 $\frac{2}{6}$ or $\frac{1}{3}$ **2** $\frac{8}{9}$ **3** $\frac{17}{12}$ or $1\frac{5}{12}$ **4 a** 10 **b** −2 **c** $\frac{1}{2}$

5 $x = 5$ **6 a** 10, 12 **b** 19, 23 **c** 18, 24 **d** 18, 15 **e** 36, 49 **f** 10 000, 100 000

7 a 1 **b** 5 **c** 9 **8 a** 2 **b** 4 **c** 8

   Module 5: Algebra and Space, shape and measure

1 $\dfrac{2}{3} \times \dfrac{1}{2} =$

2 $\dfrac{2}{3} \div \dfrac{3}{4} =$

3 $\dfrac{2}{3} + \dfrac{3}{4} =$

4 Find the value of p in each of these equations.

    a $p = 4 + 3 \times 2$ **b** $4 = p + 3 \times 2$ **c** $4 = 3 + 2 \times p$

5 Solve $4x - 7 = 13$.

6 Write the next **two** terms of each sequence.

    a 2, 4, 6, 8, …, …   **b** 3, 7, 11, 15, …, …

    c 3, 4, 6, 9, 13, …, …   **d** 30, 27, 24, 21, …, …

    e 1, 4, 9, 16, 25, …, …   **f** 1, 10, 100, 1000, …, …

7 Work out the values for the expression $4n - 3$, when:

    a $n = 1$

    b $n = 2$

    c $n = 3$

8 Work out the values for the expression 2^n, when:

    a $n = 1$

    b $n = 2$

    c $n = 3$

			Key words	
Incorporating exercise:	25A		brackets	expression
Homework:	25.1		cancel	factorise
Examples:	25.1		cross-multiply	

Learning objective(s)

- simplify algebraic fractions
- solve equations containing algebraic fractions

Prior knowledge

Pupils must have a good understanding of adding, subtracting, multiplying and dividing normal fractions. They must also be happy in factorising, multiplying and simplifying algebraic expressions.

Starter

Ask pupils to:
- Work out the following: $\frac{1}{3} + \frac{2}{5}$, $\frac{3}{4} - \frac{3}{8}$, $\frac{2}{5} \times \frac{1}{8}$, $\frac{2}{5} \div \frac{1}{8}$. $(\frac{11}{15}, \frac{3}{8}, \frac{1}{20}, 3\frac{1}{5})$

- Work these out: $6m \times 3p$, $6m \times 3m$, $6m^2 \times 3mp$, $2(3x + 5)$ $(18mp, 18m^2, 18m^3p, 6x + 10)$
  For A and A* pupils: $(3x + 5)(x - 2)$ $(3x^2 - x - 10)$
- Factorise these: $8m + 12p$, $3p^2 - 3pq$, $8ab - 4bc$ $(4(2m + 3p), 3p(p - q), 4b(2a - c))$

Main teaching points

When adding and subtracting algebraic fractions, a common denominator must be found, just as if you are adding and subtracting normal fractions. Remember, you are only changing the appearance of the fraction and not it's actual value – in effect you are multiplying each fraction by 1.

For example: $\dfrac{x}{3} + \dfrac{x}{2} = \dfrac{2}{2} \times \dfrac{x}{3} + \dfrac{3}{3} \times \dfrac{x}{2} = \dfrac{2x}{6} + \dfrac{3x}{6} = \dfrac{5x}{6}$

$\dfrac{2}{2} = 1$ $\dfrac{3}{3} = 1$ Same values as before, they just look different.

Common mistakes

For example, $\dfrac{x}{3} + \dfrac{x}{2} = \dfrac{2x}{5}$, that is pupils have no understanding.

Differentiation

All but the good grade A pupils will need quite a bit of time to be able to handle questions where factorising and cancelling are required.

Plenary

Get the pupils to multiply some algebraic terms together as practice for the homework.
For example, $3(a - 5)$, $4(2a + 1)$, $2(4 - a)$, $6(a^2 - 5a)$, $(3 + a)(a - 5)$.

$(3a - 15, 8a + 4, 8 - 2a, 6a^2 - 30a, a^2 - 2a - 15)$

Incorporating exercise:	25B	Key words	
Homework:	25.2	linear	substitute
Examples:	25.2	non-linear	

Learning objective(s)

● solve linear and non-linear simultaneous equations

Prior knowledge

Pupils should be able to solve linear simultaneous equations (see Chapter 19, Section 19.5) and be confident in algebraic manipulation.

Starter

Use the substitution method to solve this pair of simultaneous equations.

$2x + y = 5$ and $x + 3y = 5$ $(x = 2, y = 1)$

Ask the pupils to explain what steps are taken in the process of solving, and why.

Main teaching points

Always solve combinations of non-linear and linear simultaneous equations by the substitution method (not the elimination method).

Common mistakes

Solving for x (or y), and then not finding the value(s) of the other letter.

Differentiation

Lower ability pupils *must* be confident with solving linear simultaneous equations before attempting non-linear types.

Plenary

Write a pair of equations on the board, one linear, one quadratic.
For example: $y = x + 3$ and $y = x^2 + 2x + 1$
             $2y = x - 4$ and $y = x^2 - 4x + 7$
               $y + x = 1$ and $y = x^2 - 4$
Ask the pupils what should be done, in what order, and why. You don't need to actually calculate the answers, just talk through the process step by step, in order to make sure the method is clear.

Incorporating exercises: 25C, 25D
Homework: 25.3
Example: 25.3

Key words

coefficient
consecutive
difference

nth term
sequence
term

Learning objective(s)

- recognise how number sequences are built up
- find the nth term of a sequence
- recognise some special sequences

Prior knowledge

Pupils need to be able to substitute into simple algebraic expressions and have some experience in continuing a sequence.

Starter

Put the numbers 2 and 4 on the board. What is the next number in this pattern? (6 or 8 are the most common answers – discuss why we cannot be certain.)
Do the same with the numbers 1 and 5.
How many numbers do you need in a pattern before you can know what the rule is for finding the next term? Does it depend on the type of pattern?

Main teaching points

Finding the difference between terms is the first thing to do with a pattern you don't recognise. If this fails to get a result, look for another angle of attack, for example repeated multiplication if the numbers increase quickly.

Common mistakes

Only looking at the first two numbers and *assuming* the rest follow that pattern.

Plenary

Ask the pupils to make up their own number pattern and then swap books with another pupil.
"Can your neighbour work out your pattern, and can you work out theirs?" Put the difficult sequences up on the board for discussion.

Incorporating exercise:	25E		Key words	
Homework:	25.4		difference	rule
Examples:	25.4		pattern	

Learning objective(s)

● find the nth term from practical problems

Prior knowledge

Pupils need to be able to substitute into simple algebraic expressions and have some experience in continuing a sequence. Successfully completing Exercises 25C and 25D will help enormously for Exercise 25E.

Starter

Remind pupils how to find the next term and the nth term of a sequence. Give two or three for practice.
For example, 7, 9, 11, 13, 15, ...   7, 17, 27, 37, ...   7, 107, 207, 307, ...

Main teaching points

Some pupils get confused with these problems, but if they just sift through the words to get to the maths, write down the sequence, and concentrate on the numbers rather than if the question says fences or pentagons etc., they should be fine. Especially if they managed to cope with Exercises 25C and 25D.

Common mistakes

Not writing down the sequence and attempting to do the question in their head.

Differentiation

Some pupils will need help actually writing down the sequence to start with.

Plenary

Use a sequence such as the number of matches needed to make these hexagons.

 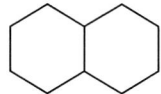 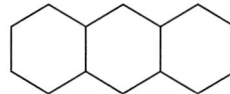

Ask the pupils to explain what steps to take, to find out how many matches would be used in the 50th shape.

Incorporating exercise:	25F		Key words	
Homework:	25.5		non-linear	second
Examples:	25.5		quadratic rule	difference

Learning objective(s)

● work out the nth term of a non-linear rule

Prior knowledge

Pupils must know the square numbers (n^2), and should be able to recognise simple changes to them, for example $n^2 - 1$, $n^2 + 1$. Finding the nth term in the form $an^2 + bn + c$ requires basic algebraic manipulation skills.

Starter

List the first 10 square numbers.
List the first five numbers of the sequences: $n^2 + 1$, $n^2 + 2$, $n^2 - 1$, $n^2 - 2$
Find the value of c in the following sequences:

c, 3, 6, 11, 18, 27		($c = 2$)
c, –1, 2, 7, 14, 23		($c = -2$)
c, 0, 3, 8, 15, 24		($c = -1$)

Main teaching points

Shown all options, many pupils tend to stick to one method – the '$an^2 + bn + c$' method, as it works for all quadratics. More able pupils will use it when the other options don't work. Pupils must be encouraged to check if the rule is actually a quadratic before any further action is taken.

Common mistakes

Getting a quadratic rule, but not calculating whether it actually works.

Differentiation

For lower ability pupils, only show them the '$an^2 + bn + c$' method as it works for all quadratics.

Plenary

Ask the class to find the rule for this sequence: 1, 3, 6, 10, 15, ... (the triangular numbers: $\frac{1}{2}n^2 + \frac{1}{2}n$).

			Key words
Incorporating exercise:	25G		subject
Homework:	25.6		transpose
Examples:	25.6		

Learning objective(s)

- change the subject of a formula where the subject occurs more than once

Prior knowledge

Good algebra skills are needed here, but this section will also help to develop them. Being able to use brackets, factorise and use algebraic fractions are all skills that are required.

Starter

Ask the class to:

- Factorise $ax + bx$, $\pi r^2 + 2\pi r$. $(x(a + b), \pi r(r + 2))$

- Factorise and simplify $\dfrac{ax + bx}{px + xy}$. $\left(\dfrac{x(a + b)}{x(p + y)} = \dfrac{a + b}{p + y} \right)$

- Find y in these equations: $y = 2 + 3 \times 6$, $2 = y + 3 \times 6$, $3 = 2 + y \times 6$, $6 = 3 + 2 \times y$
  $(y = 20, y = -16, y = \frac{1}{6}, y = 1\frac{1}{2})$

Main teaching points

With problems such as find y when $y = 2 + 3 \times 6$, we are looking for the straightforward calculation of the expression, and so we use BODMAS. However, when the problem is find y when $3 = 2 + y \times 6$, we need to work backwards (SAMDOB), and this is the process for changing the subject. It is very similar to solving an equation, but instead of a numerical answer, we get an algebraic one.

Common mistakes

Not doing the *opposite*, for example, $a + b = c$, and wrongly writing $a = c + b$.

Not using brackets or a clear division, for example, for $2a + b = c$, wrongly writing $a = c - b \div 2$, instead of either $a = (c - b) \div 2$, or $a = \dfrac{c - b}{2}$.

Plenary

Ask the pupils what steps need to be taken to make x the subject of the following:

$3p = x + E$, $3x + p = x - y$, $4(x + 2y) = 2(y + x)$, $t = \dfrac{xy}{x + y}$

Inequalities and regions

Overview

26.1 Solving inequalities
26.2 Graphical inequalities
26.3 Problem solving

This chapter takes the pupils from algebraic inequalities in Section 26.1, to displaying inequalities graphically in Section 26.2. Section 26.3 gives a whole range (grades D to A) of problems to solve by using and applying inequalities.

Context

Inequalities are used to maximise profits and minimise wastage in a whole range of industrial situations. They are also used in economics to help forecast growth, profit and loss, etc.

AQA B references

AO2 Number and algebra: Equations, formulae and identities

26.1, 26.2   2.5e "solve linear inequalities in one variable, and represent the solution set on a number line; solve several linear inequalities in two variables and find the solution set"

Route mapping

Exercise	D	C	B	A	A*
A		1–3	4–6		
B		1–3	4		
C			all		
D		1–4	5–12		
E	1	2	3–6	7–10	

Answers to diagnostic Check-in test

1 a $x = 3$ **b** $x = 18$ **c** $x = 3$ **d** $x = 2$

2

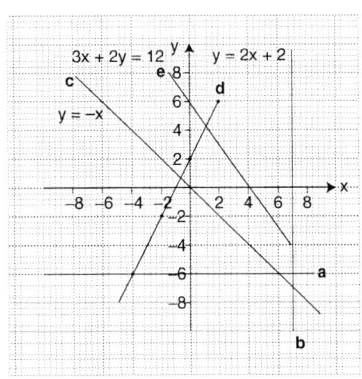

              Module 5: Algebra and Space, shape and measure

1 Solve the following equations.

   a $4x + 5 = 17$ **b** $\frac{x}{3} - 2 = 4$

   c $2x + 3 = x + 6$ **d** $3(x + 4) = 2(4x + 1)$

2 On the axes below, draw the graphs of the following.

   a $y = -6$ **b** $x = 7$ **c** $y = -x$
   d $y = 2x + 2$ **e** $3x + 2y = 12$

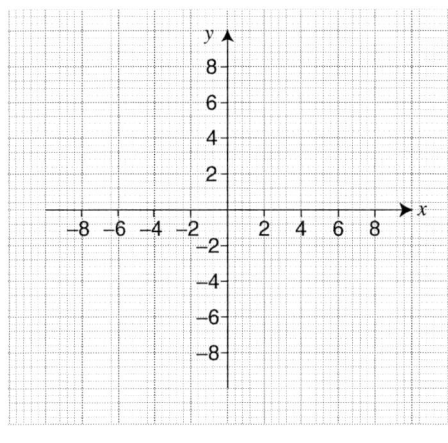

Incorporating exercises:	26A, 26B, 26C
Homework:	26.1
Examples:	26.1

Key words
inequality
number line

Learning objective(s)

● solve a simple linear inequality

Prior knowledge

Pupils must be able to solve linear and quadratic equations, including those linear equations with the variable being on both sides of the equation.

Starter

Solve for x:

● $x + 7 = 12$ ● $x - 7 = 12$

● $3x + 7 = 22$ ● $\frac{x}{3} - 7 = 22$

● $x^2 = 9$ ● $2x^2 = 72$ ● $x^2 - 1 = 15$

Main teaching points

When solving inequalities, write '$>$, $\geq$, $\leq$, or $<$' but think ' $=$ '. In other words, solve as if it is an equation, but keep the inequality signs. (Be careful when there are minus signs.)

Common mistakes

Not splitting up the inequality properly, for example for $7 < 2x + 1 < 13$ splitting it into $7 < 2x$ and $1 < 13$ rather than $7 < 2x + 1$ and $2x + 1 < 13$.

Plenary

Put a variety of inequalities on the board and ask one pupil at a time for one step of the calculation.

For example:

$x + 24 \leq 32 \ (x \leq 8)$

$2x - 1 > -6 \ (x > -2.5)$

$-3 < 4x + 1 < 17 \ (-1 < x < 4)$

$6 - 3x \geq 18 \ (x \leq -4)$

Incorporating exercise:	26D
Homework:	26.2
Examples:	26.2

Key words
boundary   origin
included   region

Learning objective(s)

● show a graphical inequality and how to find regions that satisfy more than one graphical inequality

Prior knowledge

Pupils must be able to draw graphs such as, $x = 2$, $y = -4$, $y = 2x + 2$, $6x - 8y = 24$, $y = x$.

Starter

Ask the pupils to draw a set of axes with x and y values going from -8 to $+8$. Ask them to draw the lines suggested above, one at a time. After completing each one, ask a volunteer to draw the line on the board.

Main teaching points

Whatever the inequality, draw as though the inequality were an '=' sign. Then, if the sign is $>$ or $<$, draw a dotted line, and if the sign is $\leq$ or $\geq$, draw a solid line. Think carefully whether you want the region above or below the line.

Common mistakes

Not drawing the correct line, or mixing up whether the line should be dotted or solid. Choosing the incorrect side of the line as the one required. Testing a point to check, helps to eliminate these mistakes.

Differentiation

Even though only grades B and C are tested here, it is surprising the number of pupils that find this difficult, and so will require extra help and time.

Plenary

Give some simple inequalities, for example $x \geq 2$, $y < -3$, $x > y$, $y \leq 2x + 1$, and ask some volunteers to explain the steps to take while you work at the board.

Incorporating exercise:	26E
Homework:	26.3
Examples:	26.3

Learning objective(s)

● use inequalities to solve problems

Prior knowledge

Pupils will need to have successfully completed Exercises 26A, 26B, 26C and 26D.

Starter

Sketch and shade the inequality $x + y \geq 30$.
If x represents girls and y represents boys:
a Explain what the inequality could mean.
b Give some numbers which satisfy the inequality.
If x and y represents boys and girls in a disco where the venue holds a maximum of 200 guests, write down another inequality that must be true.
Why can't we have $x = 250$ and $y = -100$ even though these satisfy both inequalities?

Main teaching points

When attempting the more complex inequality questions, most pupils find it helpful to keep substituting in real numbers to see if they are on the right track – most pupils will not get past this method and be able to think more abstractly.

Differentiation

In Exercise 26E, some pupils will only be able to cope with questions 1 to 6 without having significant help.

Plenary

Choose any question from Exercise 26E questions 5 to 9, and ask the pupils to *explain* what one has to do and *why*.

Overview

27.1 Transformations of the graph $y = f(x)$

This chapter will show pupils how to transform and recognise graphs of various functions.

Context

Used in science to compare results of experiments.

AQA B references

AO2 Number and algebra: Sequences, functions and graphs

27.1   2.6g "apply to the graph of $y = f(x)$ the transformations $y = f(x) + a$, $y = f(ax)$, $y = f(x + a)$, $y = af(x)$ for linear, quadratic, sine and cosine functions $f(x)$"

Route mapping

Exercise	D	C	B	A	A*
A					all

Answers to diagnostic Check-in test

1 a $\vec{AB} = \begin{pmatrix} 0 \\ 3 \end{pmatrix}$ **b** $\vec{AC} = \begin{pmatrix} 2 \\ 2 \end{pmatrix}$ **c** $\vec{AD} = \begin{pmatrix} 3 \\ 0 \end{pmatrix}$

2 a 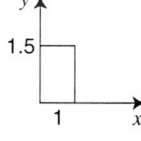   **b** 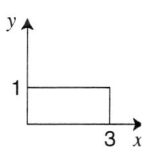   **3 a** 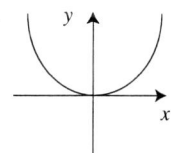   **b**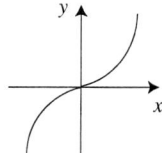

c 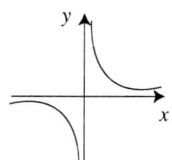   **d** 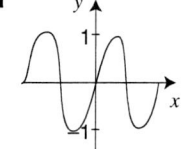   **e** 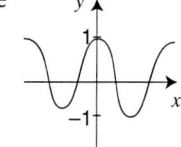   **f**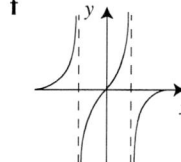

1 Write the column vector for each of the following.

1 a $\vec{AB} = \begin{pmatrix} \\ \end{pmatrix}$

  b $\vec{AC} = \begin{pmatrix} \\ \end{pmatrix}$

  c $\vec{AD} = \begin{pmatrix} \\ \end{pmatrix}$

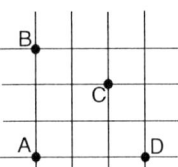

2 On the axes given, stretch the original graph by:
  a a scale factor of 1.5 in the y-direction,
  b a scale factor of 3 in the x-direction.

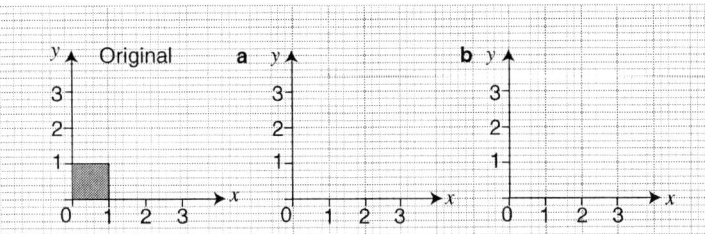

3 On each set of axes, sketch the graphs of the following.

  a $y = x^2$

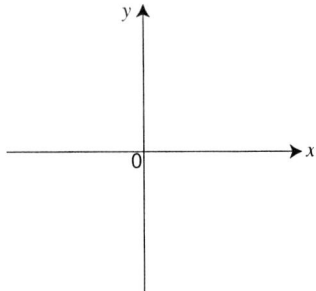

  b $y = x^3$

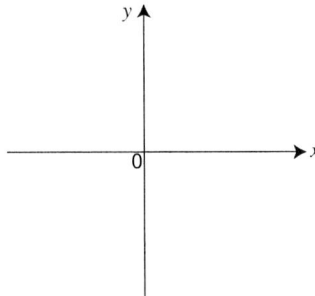

  c $y = \dfrac{1}{x}$

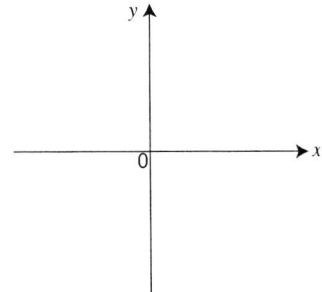

  d $y = \sin x$

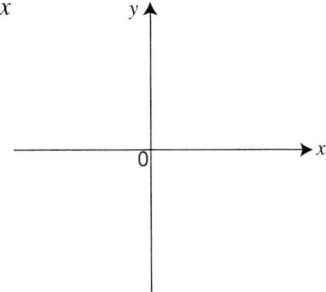

  e $y = \cos x$

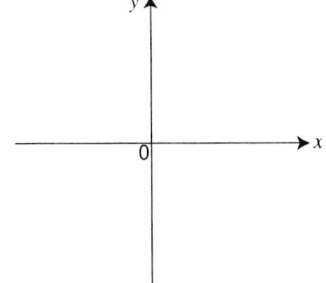

  f $y = \tan x$

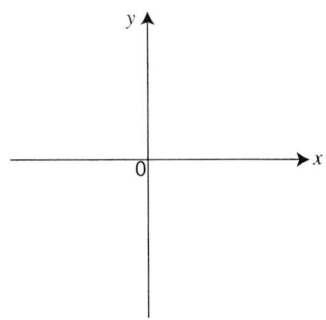

      Module 5: Algebra and Space, shape and measure

			Key words	
Incorporating exercise:	27A		function	transform
Homework:	27.1		reflection	translation
Examples:	27.1		scale factor	vector
			stretch	

Learning objective(s)

● be able to transform a graph

Prior knowledge

Pupils need to remember the basics of transforming a shape by a translation (as a vector), reflection, and enlargement (stretching in one direction only).

They also need to be able to sketch the graphs of:

$y = \frac{1}{x}$, $y = x^2$, $y = x^3$, $y = \sin x$, $y = \cos x$, and $y = \tan x$

Starter

Ask the pupils to sketch the graphs of the functions given in the Prior knowledge section.
Remind (or show) them how to stretch a graph.
Remind them of what happens when we move a point by using a column vector such as:

$$\begin{pmatrix} 2 \\ 0 \end{pmatrix}, \begin{pmatrix} 3 \\ 0 \end{pmatrix}, \begin{pmatrix} -2 \\ 0 \end{pmatrix}, \begin{pmatrix} 0 \\ -3 \end{pmatrix}.$$

Main teaching points

Make sure that your pupils can sketch all of the graphs above. This A* topic, if broken down, is accessible to A grade pupils, but there is a lot of confusing similarity between the different transformations, which only the best tend to get to grips with easily.

Common mistakes

Misunderstanding of the various rules.

Differentiation

A true A* topic. Many good pupils will get GCSE marks for sketching the graphs, but will fall down on describing the transformations.

Plenary

Sketch the graph of any function on the board. For example:

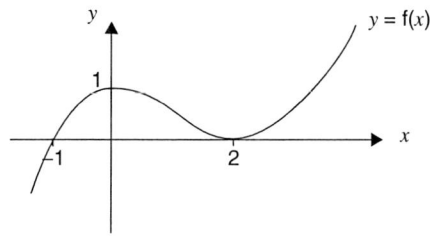

Ask what would happen to this graph if it underwent the following transformations:
$y = 2f(x)$, $y = f(x - 2)$, $y = f(x + 1)$, $y = f(x) + 1$, $y = -f(x)$

Overview

28.1 Proving standard results
28.2 Algebraic proof

This chapter shows the standard GCSE proofs that pupils *must know*, and then leads to a variety of algebraic proofs, which the pupils must be able to understand and use.

Context

Pupils who want to pursue careers in scientific research and statistics will need to prove ideas and hypotheses, so this chapter gives a basic idea of the processes involved.

AQA B references

AO2 Number and algebra: Using and applying number and algebra

28.1, 28.2 2.1j "... understanding the importance of a counter-example ..."
2.1k "understand the difference between a practical demonstration and a proof"
2.1l "show step-by-step deduction in solving a problem; derive proofs using short chains of deductive reasoning"

Route mapping

Exercise	D	C	B	A	A*
A					all
B					all
C					all

Answers to diagnostic Check-in test

1 a 70°   **b** Isosceles – 2 equal angles

2 a Any 5 examples   **b** No, because a proof must show that it is true for *all* numbers.

3 ∠AXB = 110° (angles in a triangle), so ∠DXC = 110° (vertically opposite angles)

∠XDC = 30° (alternate angles) and ∠XCD = 40° (alternate angles)

so the angles are the same and since it is a parallelogram AB = DC and AX = XC and DX = XB

so triangle ABX is congruent to triangle CDX

1 a Find *x*.

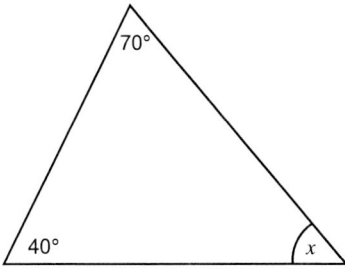

b What type of triangle is this? How do you know?

2 The product of two odd numbers is always odd.
a Give **five** examples to show this is true.

b Is this enough examples to *prove* the statement is true? Why?

3 ABCD is a parallelogram. X is the point where the diagonals meet.

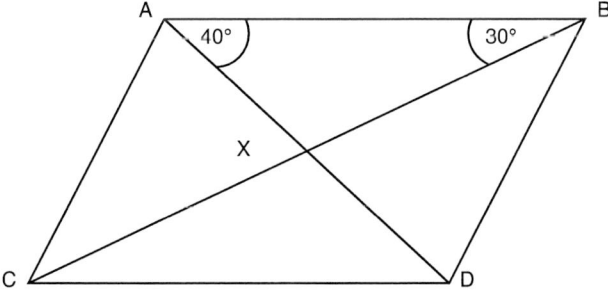

Show that triangle ABX and triangle CDX are congruent.

Incorporating exercise:	28A
Homework:	28.1

Key words

demonstration	prove
proof	show that

Learning objective(s)

● understand the difference between a proof and a demonstration

Prior knowledge

All the previous chapters provide the ground work for being able to produce formal proofs.

Starter

Ask the following questions:
● What do the angles in a triangle add up to?
● What is Pythagoras' theorem?
● What sort of number do you get when you add two odd numbers together? Give some examples.
● Give any three consecutive numbers, for example, 4, 5, and 6. How can we write them algebraically? Answer: n, $n + 1$, $n + 2$
● Find a and b. What do you notice?

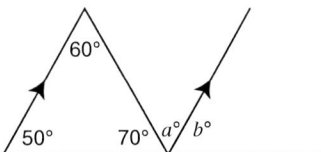

Main teaching points

The three main proofs (on pages 596 to 599 of the Pupil Book) as well as one of the congruency proofs can be learned by all good pupils. Proofs that rely on logic require more mathematical ability.

Common mistakes

Not learning the main proofs.

Differentiation

Most higher tier pupils can learn the standard proofs, but it does require logical A* pupils to think their way through most of the questions in Exercise 28A.

Plenary

Ask, "What steps can you use to understand a question involving a proof?"
(Use numbers to start with, to help pupils understand what it is they need to prove, then tell them to try and do it using algebra.)

Incorporating exercises: 28B, 28C

Homework: 28.2

Key words
prove   verify
show

Learning objective(s)

● give a rigorous and logical algebraic proof

Prior knowledge

All the previous chapters provide the ground work for being able to produce formal proofs.

Starter

Ask pupils to expand and simplify the following:

$n^2 + (n + 1)^2 - (n + 2)^2$ (Answer: $n^2 - 2n - 3$)

$(a + b)^2 + (a - b)^2$ (Answer: $2a^2 + 2b^2$)

$(n - 2)(n - 1)(n)(n + 1)$ (Answer: $n^4 - 2n^3 - n^2 + 2n$)

Main teaching points

If pupils can cope with Exercise 28A, most of Exercise 28B is a logical follow on. Exercise 28C however, is a good time filler for the mathematically inquisitive, but remember the phrases: "you can't say 'yes' or 'no' without a reason" and "20 examples is not enough to *prove* it" – sit back and enjoy your coffee!

Common mistakes

Confusing verification with proof.

Plenary

Think of a number, double it, add 2, halve it, subtract 1, what do you end up with? Hopefully, the number you first started with. Verify this works, prove this works.

Chapter 1

Homework 1.1

1 a Data is not quantitative **b** Data has an outlier
  c The mode does not use all the values. There may also not be a mode in this case.
2 a Mode 31, median 27.5, mean 23.7
  b Ritsa's sales in January (2) is a clear outlier. This low figure may not be Ritsa's fault and could be due to transport strikes, terrorist threats, etc.
3 37   **4** 9

Homework 1.2

1 a 1   **b** 3   **c** 3.32
2 a £8.95   **b** £8.95   **c** £8.75   **d** £8.00

Homework 1.3

1 a $140 \leq x < 160$ **b** 140.3 g   **c** 13.8%
2 a $6 \leq x < 8$ **b** 7.36 cm

Homework 1.4

1 b 2.7 pipes fitted
2 b   **c** 43.73 years

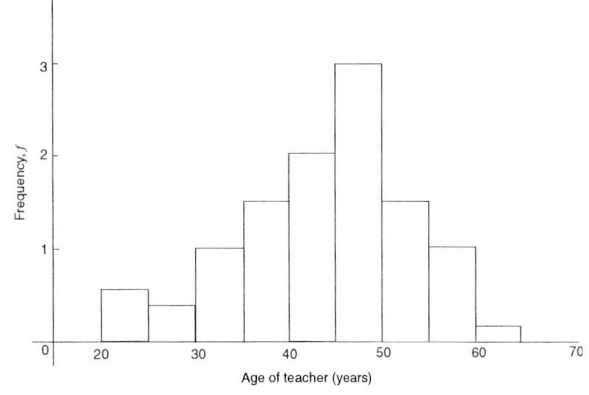

Homework 1.5

1 a   **b** 38.1 cm   **c** 36.25 cm

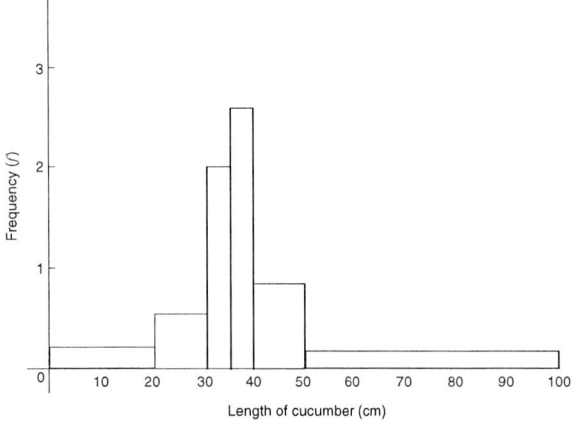

  d Lower quartile 31 cm, upper quartile 42.5 cm, interquartile range 11.5 cm

Homework 1.6

1 a

Period	1	2	3	4	5	6	7	8
Moving average	39	37.25	33.75	32	31.75	29.75	29.25	28.75

b

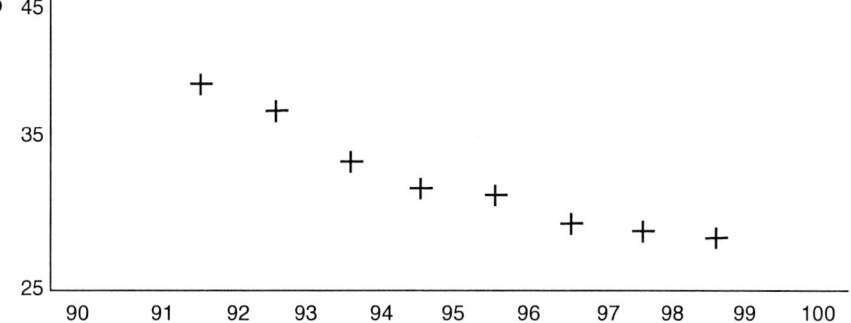

c The moving averages indicate a clear downward trend in the percentage required to gain a grade C in GCSE Mathematics which appears to be flattening off. Possible reasons left for discussion!

2 a

Day	1–6	2–7	3–8	4–9	5–10	6–11	7–12	8–13	9–14	10–15	11–16	12–17	13–18
Moving average	195	187	179	180	182	175	177	183	196	202	214	222	227

b Initially, the audience size is generally decreasing. During the second and third weeks of the season, the audience size has an upward trend.

Homework 1.7

1 Variety of answers possible. Ensure a single, simple and appropriate question, with all possible responses catered for, and space for tallying answers.

2 Variety of answers possible. Questionnaires should have two questions, relevant to the matter in question, following the guidelines in the Pupil Book, with response options which are unambiguous and cater for all eventualities.

Homework 1.8

1 Yes, because the lowest level of support is 51%.

2 a

Year	1980	1985	1990	1995	2000	2005
Index	100	108	115	119	122	130
Price	26p	28p	30p	31p	32p	34p

b 1980–1985

3 a 1990–2000   **b** 1950–60   **c** 130 million

Homework 1.9

1 Numbers in each respective age group are calculated to be 9, 6, 3, 7, 14, 12 (rounded to the nearest whole number). However, this totals to 51 not 50. Therefore, pupils must subtract 1 from any age group.

2 Allow any sensible responses. Possible advantages include: generally gives more accurate results; can focus on important "subpopulations" such as particular ethnic groups or areas of the country; less "relevant" subpopulations can be ignored. Possible disadvantages include: can be difficult to select appropriate strata; can be expensive; requires accurate initial information to determine size of subpopulations.

Homework 2.1

1 a Day 7, 65 m   **b** Day 35, 26 m   **c** About 59 m

2 a   **b** 42°C   **c** About 16°C

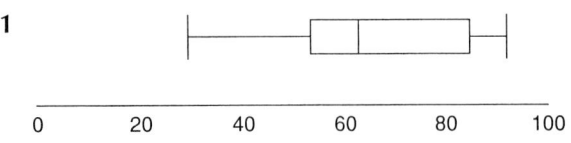

Homework 2.2

1 a 32, 32, 37, 37, 38, 38, 38, 41, 41, 44, 44, 44, 46, 46, 47, 47, 48, 50, 53, 53

b

Key	3		2 represents £32
6	2 2 7 7 6 8 8
4	1 1 4 4 4 6 6 6 7
5	0 3 3

c £40 to £50 range, because the 4 stem has the most leaves

2 a 18 months   **b** £48   **c** £31

Homework 2.3

1 a Positive   **c** About 42 m (depending on line of best fit)   **d** About 1.69–1.70 m (depending on line of best fit)

2 b Negative **d** About £1800 to £1900 (depending on line of best fit)
e About 40 000 miles (depending on line of best fit)

Homework 2.4

1 a 2, 7, 17, 20   **b** Points plotted should be (4, 2), (7, 7), (10, 17), (13, 20)   **c** About 8 cm
d About 3 cm   **e** The interquartile range is only 3 cm, so the heights are very consistent

2 a 3, 7, 13, 25, 30   **b** Points plotted should be (20, 3), (40, 7), (60, 13), (80, 25), (100, 30)   **c** About 63%
d Approximately 22 pupils would pass

Homework 2.5

1

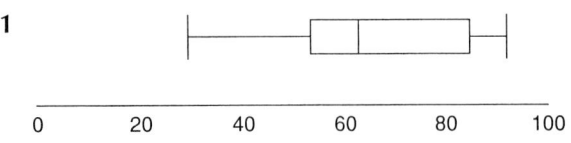

2

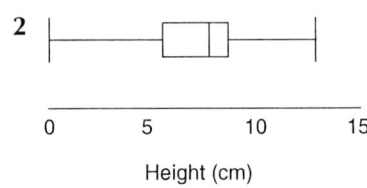

3 The median time for Lastalot is 360 hours, while for Everbrite it is 330 hours. However, although Everbrite's median times are slightly lower, the consistency is much greater as the interquartile range is 80 compared to Lastalot's 250. Everbrite's lightbulbs would most likely be the best buy.

Homework 2.6

1 a $\bar{x} = 2.5$, $\sigma = 1.12$ **b** $\bar{x} = 15$, $\sigma = 3.42$ **c** $\bar{x} = 6.975$, $\sigma = 0.904$

2 a 32.5, 37.5, 45, 55   **b** $\sigma = 10.1$ cm

    c Victor's marrows have a higher mean and smaller standard deviation than Lucy's. This means that, on average, not only are his marrows larger but that their lengths are more consistent.

Chapter 3

Homework 3.1

1 a White $\dfrac{7}{20}$, red $\dfrac{17}{120}$, yellow $\dfrac{3}{10}$, blue $\dfrac{5}{24}$ **b** White 7, red 3, yellow 6, blue 4

2 a 0.10, 0.09, 0.11, 0.10, 0.09, 0.11, 0.11, 0.10, 0.09, 0.09 (all to two decimal places)
   b Results are fairly close, but you might expect them to be a little closer after 1500 trials, so may require more evidence.

3 a $\dfrac{13}{20}$, $\dfrac{19}{25}$, $\dfrac{31}{50}$, $\dfrac{33}{50}$, $\dfrac{297}{500}$ or 0.594   **b** 11 or 12 times

Homework 3.2

1 $\dfrac{2}{9}$ **2** $\dfrac{1}{10}$

3 a D/F: mutually exclusive, since a number cannot be a multiple of 7 and a factor of 50, but not exhaustive since, for example, the number 6 is not included in either event
   b A/B: a number cannot be both odd and even at the same time, so mutually exclusive and exhaustive, since a number must either be odd or even
   c The number 49 is both a square and a multiple of 7
   d E/F: the number 5 is both prime and a factor of 50
   e The number is not prime

4 No, mutually exclusive events are not independent. Consider events A and B being mutually exclusive. If A occurs, then P(B) = 0. So A occurring affects P(B).

Homework 3.3

1 a $\dfrac{2}{5}$ **b** 120   **c** 240

2 a i $\dfrac{3}{5}$ **ii** $\dfrac{1}{100}$ **iii** $\dfrac{2}{5}$

   b $\dfrac{1}{4}$ **c** 100   **d** £50

Homework 3.4

1 a 75   **b** 44%   **c** 140

2 a 28%   **b** 14%   **c** $\dfrac{27}{100}$ **d** $\dfrac{69}{100}$ **e** 2400

Homework 3.5

1 a $\dfrac{1}{3}$ **b** $\dfrac{4}{15}$ **c** $\dfrac{3}{5}$

2 a $\dfrac{3}{10}$ **b** $\dfrac{3}{10}$ **c** $\dfrac{3}{5}$

3 a $\dfrac{1}{12}$ **b** $\dfrac{1}{8}$ **c** $\dfrac{19}{24}$

4 Events are not mutually exclusive (it is possible to be stopped by both on his way to work) and therefore probabilities cannot be added

5 a 35% **b** 40% **c** 53%

Homework 3.6

1 a $\dfrac{7}{18}$ **b** $\dfrac{1}{12}$ **c** $\dfrac{7}{18}$

2 a $\dfrac{1}{4}$ **b** $\dfrac{1}{3}$ **c** $\dfrac{5}{12}$

Homework 3.7

1 a

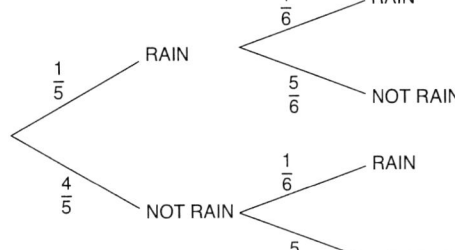

	DAY 1	DAY 2	OUTCOME	PROBABILITIES
		RAIN	RR	$\frac{1}{5} \times \frac{1}{6} = \frac{1}{30}$
	RAIN	NOT RAIN	RN	$\frac{1}{5} \times \frac{5}{6} = \frac{5}{30} = \frac{1}{6}$
		RAIN	NR	$\frac{4}{5} \times \frac{1}{6} = \frac{4}{30} = \frac{2}{15}$
	NOT RAIN	NOT RAIN	NN	$\frac{4}{5} \times \frac{5}{6} = \frac{20}{30} = \frac{2}{3}$

b $\dfrac{1}{30}$ **c** $\dfrac{1}{3}$

2 a

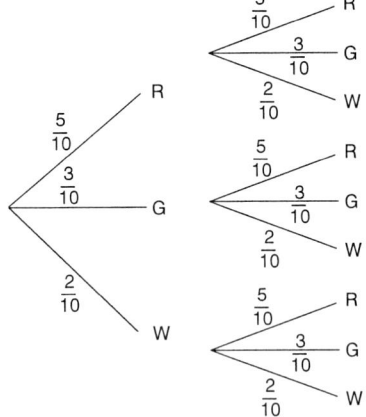

	FIRST BALL	SECOND BALL	OUTCOME	PROBABILITIES
		R	RR	$\frac{5}{10} \times \frac{5}{10} = \frac{25}{100}$
		G	RG	$\frac{5}{10} \times \frac{3}{10} = \frac{15}{100}$
	R	W	RW	$\frac{5}{10} \times \frac{2}{10} = \frac{10}{100}$
		R	GR	$\frac{3}{10} \times \frac{5}{10} = \frac{15}{100}$
	G	G	GG	$\frac{3}{10} \times \frac{3}{10} = \frac{9}{100}$
		W	GW	$\frac{3}{10} \times \frac{2}{10} = \frac{6}{100}$
		R	WR	$\frac{2}{10} \times \frac{5}{10} = \frac{10}{100}$
	W	G	WG	$\frac{2}{10} \times \frac{3}{10} = \frac{6}{100}$
		W	WW	$\frac{2}{10} \times \frac{2}{10} = \frac{4}{100}$

b $\dfrac{9}{100}$ **c** $\dfrac{19}{50}$ **d** $\dfrac{3}{4}$

Homework 3.8

1 a $\dfrac{1}{16}$ **b** $\dfrac{3}{8}$ **c** $\dfrac{9}{16}$

2 a $\dfrac{1}{10}$ **b** $\dfrac{1}{20}$ **c** $\dfrac{39}{40}$

3 a $\dfrac{1}{8}$ **b** $\dfrac{1000}{2197}$ **c** $\dfrac{433}{2197}$ **d** $\dfrac{37}{64}$

4 a $\dfrac{1}{2^n}$ **b** $1-\dfrac{1}{2^n}$ **c** $\dfrac{n}{2^n}$

Homework 3.9

1 a $\dfrac{3}{28}$ **b** $\dfrac{15}{28}$ **c** $\dfrac{5}{14}$ **2 a** $\dfrac{2}{91}$ **b** $\dfrac{20}{91}$

3 a $\dfrac{11}{850}$ **b** $\dfrac{1}{5525}$ **c** $\dfrac{22}{425}$ **4** $\dfrac{53}{95}$

5 a $\dfrac{495}{500}\times\dfrac{494}{499}\times\dfrac{493}{498}=\dfrac{33}{100}\times\dfrac{247}{499}\times\dfrac{493}{83}=\dfrac{4018443}{4141700}=0.97$

    b $\dfrac{495}{500}\times\dfrac{5}{499}\times\dfrac{494}{498}=\dfrac{165}{100}\times\dfrac{1}{499}\times\dfrac{247}{83}=\dfrac{40755}{4141700}=0.00984$

Chapter 4

Homework 4.1

1 a 4320   **b** £2.88   **2** £330   **3 a** £54   **b** 10%   **4** 13

Homework 4.2

1 21   **2** 8.9   **3** 18.6   **4** 240   **5** 2.62
6 280   **7** 0.18   **8** 220   **9** 750   **10** 4

Homework 4.3a

1 a 60 000   **b** 60 000   **c** 40   **d** 100   **e** 100
   f 0.01   **g** 0.1   **h** 0.0007

2 a 10 000   **b** 10 000   **c** 9970   **d** 9969   **e** 9968.5
   f 9968.51

3 a 850 to 949   **b** 1650 to 1749   **c** 31 450 to 31 549
   d 950 000 to 1 499 999

Homework 4.3b

1 a 150 000   **b** 15   **c** 150   **d** 0.15   **e** 45 000
   f 50   **g** 0.75   **h** 1200   **i** 2 000 000   **j** 3 000 000

2 a 35 000   **b** 105   **c** 500   **d** 4200   **e** 100
   f 25

3 a £7.14   **b** 6 million or 6.3 million   **c** £19 000 or £20 000
   d 43 mph   **e** You don't round a phone number!   **f** 1000 km or 1100 km

Homework 4.4

1 3, 6, 9, 12, 15, 18   **2** 6, 12, 18   **3** 1, 2, 3, 6, 9, 18   **4** 2, 3, 5, 7, 11, 13, 17, 19, 23, 29
5 1, 4, 9, 16, 25, 36, 49, 64, 81, 100, 121, 144, 169, 196, 225   **6** 1, 8, 27, 64, 125
7 ±7   **8** ±9   **9** 5   **10** 4
11 1, 3, 6, 10, 15, 21, 28   **12** 0.7   **13** 40 seconds   **14** 5.5

Homework 4.5

1 a $2^3 \times 3^2 \times 5$ **b** $2^3 \times 13$ **2** 56   **3** 756   **4** 12

Homework 4.6

1 a −18   **b** 20   **c** −18   **d** 12   **e** 5
   f −12   **g** −5   **h** −7   **i** 6   **j** −2
   k −10

2 −39   **3** $5 + (-4 - -3) \times 2 = 3$

Chapter 5

Homework 5.1

1 $\dfrac{3}{10}$ **2** $\dfrac{360}{2000} = \dfrac{9}{50}$ **3** $\dfrac{45}{180} = \dfrac{1}{4}$ **4** $1\dfrac{200}{1000} = 1\dfrac{1}{5}$

5 $\dfrac{150}{250} = \dfrac{3}{5}$ **6** $\dfrac{12}{54} = \dfrac{2}{9}$ **7** $\dfrac{16}{40} = \dfrac{2}{5}$ **8** $\dfrac{8}{80} = \dfrac{1}{10}$

9 $\dfrac{450}{1200} = \dfrac{3}{8}$ **10** $\dfrac{12}{36} = \dfrac{1}{3}$

Homework 5.2

1 $\dfrac{9}{20}$ **2** $\dfrac{11}{30}$ **3** $\dfrac{11}{15}$ **4** $1\dfrac{5}{12}$ **5** $1\dfrac{1}{4}$

6 $\dfrac{7}{20}$ **7** $\dfrac{1}{6}$ **8** $3\dfrac{3}{4}$ **9** $\dfrac{3}{4}$ **10** $\dfrac{13}{42}$

11 $3\dfrac{1}{56}$

Homework 5.3

1 a $\dfrac{3}{5}$ **b** $\dfrac{3}{14}$ **c** $\dfrac{5}{9}$ **d** $\dfrac{1}{4}$

2 a $1\dfrac{3}{10}$ **b** $2\dfrac{1}{2}$ **c** $7\dfrac{5}{16}$ **d** $3\dfrac{3}{8}$

3 220 yards   **4** $\dfrac{3}{5}$ **5** $\dfrac{1}{4}$

Homework 5.4

1 a $\frac{3}{5}$ **b** 2   **c** $2\frac{18}{55}$ **d** $20\frac{5}{9}$ **e** $10\frac{2}{3}$

f 3   **2** 13 rooms, $1\frac{1}{4}$ m left over   **3** $1\frac{8}{9}$

Homework 5.5

1 a 1.02   **b** 1.22   **2 a** 0.98   **b** 0.78   **3** £145.60

4 35.2 kg   **5** £33 637.50   **6** £17 000

7 $A \times 0.5 \times 0.5 = A \times 0.25 = A \times 25\%$, 25% of the original is a 75% decrease

Homework 5.6

1 a 16%   **b** 42%   **c** 0.69%   **d** 12.5%

2 48%   **3** 12.0%

4 8.2%   **5** 71%

Homework 5.7

1 a £595.51   **b** £669.11   **c** £895.42

2 a £10 072.50   **b** £7277.38   **c** £4469.22

3 10 years

4 $1.05 \times 1.05 = 1.1025$ which is a 10.25% increase

Homework 5.8

1 1250 g   **2** 128 people   **3** £140   **4** £23 456   **5** £1234
6 None

Chapter 6

Homework 6.1

6.1a

1 $\frac{3}{5}$ **2** $\frac{5}{11}$ **3** $\frac{10}{19}, \frac{4}{19}, \frac{5}{19}$ **4** $\frac{5}{9}$

5 Labour $\frac{1}{3}$ Conservative $\frac{5}{9}$ Lib Dem $\frac{1}{9}$ **6** $\frac{9}{11}$

6.1b

1 £20, £30   **2** 405 g, 270 g   **3** £750   **4** 1370 votes

5 a 1 : 400 000   **b** 4.8 km   **c** 3 cm   **6 a** 1 : 1.8   **b** 1 : 9.6

6.1c

1 20 bungalows   **2** 360 girls   **3** 32 litres   **4** 10 cm   **5** 18 buckets
6 4 kg   **7** 15 cm   **8** 20 pens

Homework 6.2

1 55 mph   **2** 14 miles   **3** 71.4 mph   **4** 6 hrs 40mins   **5** 11 km/h

6 a 165 km/h   **b** 50 km   **c** 150 km/h

Homework 6.3

1 6720 boxes   **2** £18.20   **3** £220   **4** 98 tiles

5 £8.33   **6** £135   **7** £6.72   **8** 75 eggs

9 560 g flour, 105 g butter, 70 g sugar, 140 g sultanas, 385 ml milk

Homework 6.4

1 The cheapest box is the two 750 g Kellogg's at Tesco.

2 Sainsbury own brand 500 g on offer

3 250 g – 66p, 500 g – 81p, 750 g – 88p, 1 kg – 83p

Homework 6.5

1 1 g/cm^3 **2** 0.5$\dot{5}$ g/cm^3 **3** 1.12 g/cm^3

4 260 cm^3 **5** 1012.5 g   **6** 1150 cm^3

Chapter 7

Homework 7.1

1 $8^4 = 4096$ **2** $2 \times 2 \times 2 \times 2 \times 2 \times 2 \times 2 \times 2 = 256$

3 a 17   **b** 1   **c** 1

4 a $\dfrac{1}{4^2} = \dfrac{1}{16}$ **b** $\dfrac{2}{b^4}$

5 a t^{-3} **b** $3b^{-1}$

6 0.156 25   **7 a** 3^{-2} **b** 3^6 **c** 3^{-8}

8 a $12a^5$ **b** $3a^{-1}$ **c** $3ab^{-1}$

9 a 0.25 or $\dfrac{1}{4}$ **b** 0.00 001 or $\dfrac{1}{10\,000}$

Homework 7.2

1 a 56   **b** 560   **c** 0.56   **d** 0.0056

   e 56   **f** 56 565   **g** 0.565 65   **h** 0.000 56

2 a 650   **b** 0.0065

3 a 5.6×10^{-5} **b** 5.6×10 **c** 5.6×10^4

4 a 8×10^9 **b** 1.05×10^{-2} **c** 3×10^2 **d** 2×10^4

5 a 4.5×10^5 **b** 3.6×10^{-4} **c** 6.9×10^{-11}

Homework 7.3

1 a $0.\dot{6}$ **b** 0.75 **c** 0.8 **d** $0.8\dot{3}$ **e** $0.\dot{8}5714\dot{2}$

  f 0.875 **g** $0.\dot{8}$ **h** 0.9

2 a $\dfrac{2}{5}$ **b** $\dfrac{33}{40}$ **c** $\dfrac{19}{100}$ **d** $\dfrac{3}{500}$

3 a $\dfrac{7}{9}$ **b** $\dfrac{7}{90}$ **c** $\dfrac{7}{99}$

4 a 0.0625 **b** 0.01 **c** 2.5 **d** 25

Homework 7.4

1 a $\sqrt{15}$ **b** 7 **c** $\sqrt{\dfrac{3}{7}}$

2 a 4 **b** $2\sqrt{6}$ **c** $5\sqrt{2}$

  d 15 **e** $\sqrt{6}$ **f** 1

3 a $12\sqrt{6}$ **b** $24\sqrt{3}$ **c** $5\sqrt{5}$ **d** 4 **e** efg   **f** $\dfrac{5}{9}$

4 a $5 + 3\sqrt{3}$ **b** $2\sqrt{5} - 5$ **c** $5\sqrt{5} - 24$

5 a $\dfrac{\sqrt{5}}{5}$ **b** $2\sqrt{3}$ **c** $\dfrac{\sqrt{3}}{3} + 1$ **d** $-\dfrac{6}{5}\left(1 + \sqrt{6}\right)$

6 Area $= 20 + 8\sqrt{5}$ or $4(5 + 2\sqrt{5})$ Perimeter $= 16 + 4\sqrt{5}$ or $4(4 + \sqrt{5})$

Chapter 8

Homework 8.1

1 a 8 and 18   **b**   **c** 4.5

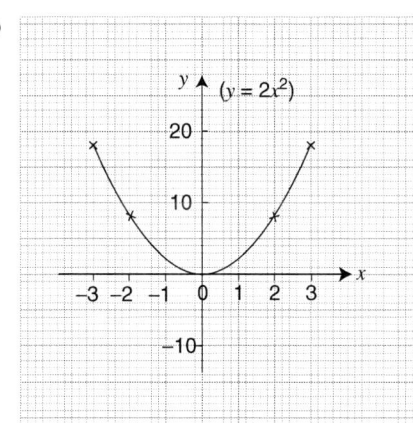

2 a −3 and 17   **b**   **c** −1.7, 1.2   **d** −1.7, 1.2

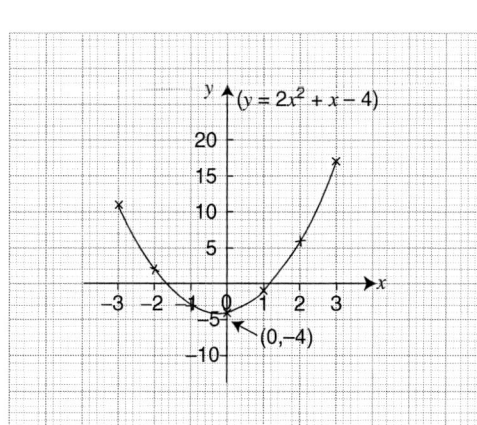

  e −2.9, 2.4   **f** $y = -4$ **g** $\left(-\dfrac{1}{4}, -4\tfrac{1}{8}\right)$

Homework 8.2

1 a $x = -0.7$ and 2.7   **b** $x = -1$ and 3   **c** $x = 1$ **d** $x = -1$ and 2   **e** $x = 1$ and 2

Chapter 9

Homework 9.1

1 a 150   **b** 1.2   **2 a** 10   **b** 13.5

3 a 64   **b** 1   **4 a** 2   **b** 900

5 a £21 125   **b** 0.9

Homework 9.2

1 a 6   **b** 30   **2 a** 6.4   **b** 24

3 a 4   **b** 4   **4 a** 0.32   **b** 0.284^r

5 a £15 337   **b** 0.4

Chapter 10

Homework 10.1

1 a $3.5\,\text{m} \leq \text{length} < 4.5\,\text{m}$ **b** $24.5\,\text{ml} \leq \text{volume} < 25.5\,\text{ml}$ **c** $65\,\text{g} \leq \text{weight} < 75\,\text{g}$
d $2500\,\text{t} \leq \text{weight} < 3500\,\text{t}$ **e** $2950\,\text{t} \leq \text{weight} < 3050\,\text{t}$

2 a $3.5\,\text{m} \leq \text{length} < 4.5\,\text{m}$ **b** $24.5\,\text{ml} \leq \text{volume} < 25.5\,\text{ml}$ **c** $65\,\text{g} \leq \text{weight} < 75\,\text{g}$
d $2500 \leq \text{number} < 3500$ **e** $2950 \leq \text{number} < 3050$ **f** $7.25 \leq \text{number} < 7.35$
g $9.305 \leq \text{number} < 9.315$ **h** $99.985 \leq \text{number} < 99.995$ **i** $99.995 \leq \text{number} < 100.005$

3 a 3.5 m to 4.5 m   **b** 2950 to 3050   **c** 246.5 to 247.5   **d** 0.0195 to 0.0205

4 a 55.5 g   **b** 1090 g = 1.09 kg

Homework 10.2

1 a $137.75\,\text{cm}^2 \leq \text{area} < 162.75\,\text{cm}^2$ **b** $176.5575\,\text{cm}^2 \leq \text{area} < 179.4475\,\text{cm}^2$

2 a $115\,\text{cm} \leq \text{width} < 125\,\text{cm}$, $205\,\text{cm} \leq \text{length} < 215\,\text{cm}$ **b** $23\,575\,\text{cm}^2$ **c** 680 cm

3 17 647 000 tonnes   **4 a** $661.5\,\text{cm}^2$ **b** $857.375\,\text{cm}^3$

5 a $9.98\,\text{cm} \leq \text{length} < 10.02\,\text{cm}$ **b** $2139.3\,\text{g} \leq \text{mass} < 2261.3\,\text{g}$

Chapter 11

Homework 11.1

1 a i 19.5 cm **ii** $30.2\,\text{cm}^2$ **b i** 88.0 m **ii** $616\,\text{m}^2$ **2** 6.18 cm

3 212 revolutions

4 $796\,\text{cm}^2$

5 a 46.3 cm   **b** $127\,\text{cm}^2$ **6** 21.5%

Homework 11.2

1 a $45\,cm^2$ **b** $32.5\,cm^2$ **c** $22.5\,m^2$ **d** $354\,m^2$ **e** $66\,m^2$

2 a $16\,cm^2$ **b** $27\,m^2$ **3** $9.3\,cm$ **4** $\frac{1}{2}(a+2b+c)h$

Homework 11.3

1 a i $2.62\,cm$ **a ii** $3.93\,cm^2$ **b i** $12.6\,cm$ **b ii** $36.4\,cm^2$ **2** $41.9\,cm$

3 $5\pi\,cm^2$ **4** $183°$ **5** $242°$ **6** $21.3\,cm$

Homework 11.4

1 a i $168\,cm^2$ **ii** $120\,cm^3$ **b i** $492\,cm^2$ **ii** $504\,cm^3$ **2 a** $270\,m^3$ **b** $1500\,cm^3$

3 $28\,cm$ **4** $15.5\,cm^2$ **5** $62.4\,kg$

Homework 11.5

1 a i $141\,cm^3$ **ii** $94.2\,cm^2$ **b i** $137\,cm^3$ **ii** $110\,cm^2$ **c i** $14\,700\,cm^3$ **ii** $2360\,cm^2$

2 $2.39\,cm$ **3** $20\,m$ **4** $42.7\,kg$ **5** $1.27\,g/cm^3$

6 $\dfrac{3\pi r^2 h}{2}$

Homework 11.6

1 a $900\,cm^3$ **b** $188\,cm^3$ **c** $24\,cm^3$ **2** $25\,cm$ **3** $1.68\,g$
4 $3.14\,cm$ **5** $176\,cm^3$

Homework 11.7

1 a i $1230\,cm^3$ **ii** $704\,cm^2$ **b i** $262\,cm^3$ **ii** $215\,cm^2$

2 $536\,cm^3$ **3** $324\pi\,cm^2$ **4** $5.76\,cm$ **5** $136\pi\,cm$

6 $183\,ml$ **7** $2.89\,cm$

Homework 11.8

1 a i $1440\,cm^3$ **ii** $616\,cm^2$ **b i** $0.524\,m^3$ **ii** $3.14\,m^2$ **2** $4.57\,cm$

3 $111\,cm^3$ **4** $48\,cm$ **5** $2.88\,cm$

6 $305\,cm^2$

Chapter 12

Homework 12.1

1 h marked appropriately on side facing right-angle.

2 a $18.4\,cm$ **b** $4.7\,m$ **c** $111.1\,cm$

Homework 12.2

1 a h   **b** s   **c** h   **d** s   **e** s
2 a $28.7\,cm$ **b** $5.7\,m$ **c** $16.4\,cm$

Homework 12.3

12.3a

1 1.73 m
2 4.5 km
3 a 5.1
b 11.2
c 6.7

12.3b

1 a 49.5 cm^2
b 9.17 cm^2
c 3.9 cm^2
2 62.4 cm^2
3 Triangle A's area is 48 cm^2, triangle B's is 54.5 cm^2, so B has the largest area.
4 13.0 cm
5 2.60 cm^2

12.3c

1 a 13 cm
b 14.3 cm
2 8.66 cm
3 a 9.43 cm
b 15.3 cm
c 11.2 cm
4 10.9 cm

Homework 12.4

1 a 0.574 | **b** 0.819 | **c** 0.700 | **d** 0.940 | **e** 0.342
f 2.75 | **g** 0.999 | **h** 0.0349 | **i** 28.6
2 a 2.74 | **b** 4.33 | **c** 44.8 | **d** 9.96 | **e** 4.33
f 7.98 | **g** 2.18 | **h** 0.0524 | **i** 1.00
3 a 11.8 | **b** 25.7 | **c** 23.5 | **d** 10.1 | **e** 11.3
f 12.2 | **g** 2.00 | **h** 3.00 | **i** 1.00

Homework 12.5

1 a 20.5 | **b** 43.6 | **c** 35.3 | **d** 7.1 | **e** 81.9
f 1.1 | **g** 29.4 | **h** 29.7 | **i** 45.0
2 a 69.5 | **b** 46.4 | **c** 54.7 | **d** 82.9 | **e** 8.1
f 88.9 | **g** 60.6 | **h** 60.3 | **i** 45.0
3 a 19.3 | **b** 34.6 | **c** 30.0 | **d** 60.0 | **e** 79.5
f 88.1 | **g** 83.0 | **h** 45.0 | **i** 30.0
4 a 45.6 | **b** 44.4 | **c** 35.5 | **d** 6.4 | **e** 63.6
f 66.0 | **g** 0.0 | **h** 90.0 | **i** 60.0

Homework 12.6

1 a 6.89 cm
b 5.00 cm
c 5.07 cm
2 a 10.5 cm
b 12.7 cm
c 10.9 cm
3 a 9.23 cm
b 7.47 cm
c 11.8 cm
4 a 45.6°
b 45.6°
c 30.0°

Homework 12.7

1 a 7.31 cm
b 6.78 cm
c 14.5 cm
2 a 10.2 cm
b 7.07 cm
c 10.4 cm
3 a 3.76 cm
b 12.9 cm
c 12.7 cm
4 a 31.0°
b 52.0°
c 38.3°

Homework 12.8

1 a 11.5 cm
b 5.77 cm
c 5.60 cm
2 a 9.53 cm
b 9.10 cm
c 8.39 cm
3 a 8.96 cm
b 13.9 cm
c 1.95 cm
4 a 21.8°
b 55.0°
c 45.0°

Homework 12.9

1 a 3.48 cm
b 7.95 cm
c 6.00 cm
2 a 9.53 cm
b 6.30 cm
c 7.72 cm
3 a 11.8 cm
b 13.4 cm
c 6.88 cm
4 a 48.8°
b 40.0°
c 45.6°

Homework 12.10

12.10a

1 19.3°   **2** 6.34 m   **3** 72.9°   **4** 13.9 cm   **5** 20.7 m
6 50°, 19.2 cm, 16.1 cm

12.10b

1 107 m   **2** 150 m   **3** 57.3 m   **4** 168 m   **5** 1250 m

12.10c

1 a 80.8 km   **b** 26.3 km   **2 a** 113 km   **b** 41.0 km   **3** 124°
4 a 41.5 km   **b** 28.0 km   **c** 52.0 km   **d** 321°

12.10d

1 a 42.8°, 42.8°, 94.3°   **b** 112 cm^2 **2 a** 11.8 cm   **b** 44.9°   **c** 20.3 cm
3 a 50.2 cm^2 **b** 550 cm^2 **c** 618 cm^2 **4** 44.8 cm^2

Chapter 13

Homework 13.1

1 a 50°   **b** 25°   **c** 28°
2 a $a = 65°, b = 115°, c = 65°$ **b** $a = 100°, b = 80°, c = 100°$
   c $a = 55°, b = 125°, c = 55°$ **d** $a = 100°, b = 30°$
   e $a = 40°, b = 40°, \ c = 20°$ **f** $a = 68°, b = 60°$
3 a $x = 40°$, rhombus or parallelogram   **b** $x = 20°$, square or rectangle

Homework 13.2

1 105°   **2 a** 18°   **b** 162°
3 177°   **4 a** 24 sides   **b** 36 sides   **5** 30 sides
6 No. Each exterior angle is 70°, and so the number of sides would be 360/70= 5.14, which is not a whole number and impossible to draw.
7 52 sides

Homework 13.3

1 $x = 60°$ **2** $x = y = 90°$
3 $a = c = 57°, b = 60°, d = 63°$ **4** $a = b = 40°, c = 115°, d = 25°$
5 $a = 28°, b = 65°, c = d = 87°$ **6** $a = 56°, b = 112°$
7 $a = b = 40°, c = 20°, d = 130°$ **8** $a = 75°, b = 35°, c = 39°, d = e = 31°$

Homework 13.4

1 $x = 95°$ **2** $x = 105°$
3 $x = 40°$ **4** $a = 28°, b = 90°, c = 62°, d = 34°, e = 118°$
5 $x = 55°$ **6** $x = 110°, y = 80°$
7 $a = 60°, b = 45°$ **8** $a = 72°, b = 108°, c = 36°$ **9** $x = 60°$

Homework 13.5

1 $x = 48°$ **2** $x = 115°$ **3** $x = 18°$
4 $y = 40°, x = 50°$ **5** $x = y = 57.5°$ **6** $x = 9$ cm
7 $x = 55°$ **8** $a = e = 57.5°, b = d = 62.5°, c = 60°$ **9** $x = 14.3$ cm, $y = 18.7$ cm

Homework 13.6

1 $x = 70°$, $y = 50°$ **2** $x = y = 35°$, $z = 55°$
3 $a = 65°$, $b = 55°$, $c = 120°$ **4** $a = b = 70°$, $c = 40°$
5 $x = 56°$, $y = 68°$ **6** $a = b = 60°$
7 $x = 59°$, $y = 62°$ **8** $x = 45°$

Chapter 14

Homework 14.1

1 a Yes, RHS   **b** Yes, SSS   **c** Yes, SAS   **d** No   **e** No
f Yes, SAS   **g** No   **h** Yes, ASA   **i** Yes, SAS
2 a Triangles ABC and DEF are congruent   **b** Triangles CAB and JHG are congruent

Homework 14.2

1 a $\begin{pmatrix} 1 \\ 4 \end{pmatrix}$ **b** $\begin{pmatrix} 3 \\ -1 \end{pmatrix}$ **c** $\begin{pmatrix} -6 \\ 2 \end{pmatrix}$ **d** $\begin{pmatrix} 2 \\ -3 \end{pmatrix}$ **e** $\begin{pmatrix} 0 \\ -2 \end{pmatrix}$ **f** $\begin{pmatrix} -3 \\ 3 \end{pmatrix}$

g $\begin{pmatrix} -1 \\ -4 \end{pmatrix}$ **h** $\begin{pmatrix} -3 \\ 1 \end{pmatrix}$ **i** $\begin{pmatrix} -5 \\ 0 \end{pmatrix}$

2 $\begin{pmatrix} -1.5 \\ -2 \end{pmatrix}$

Homework 14.3

1 a 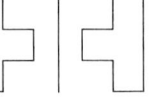   **b** 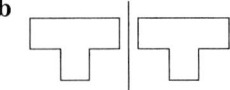   **c** 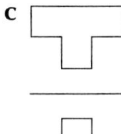   **d**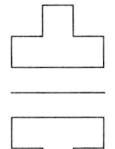

2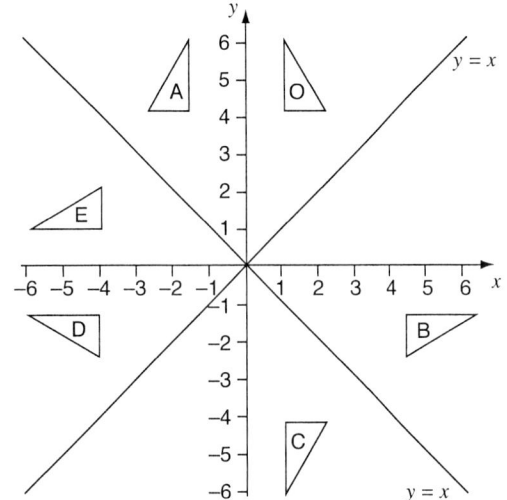

3 a (2, –5) **b** (–2, 5) **c** (0, 5) **d** (5, 2) **e** (–5, –2)
  f (2a – 2, 5)

Homework 14.4

1 a 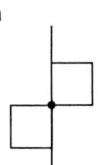 **b** 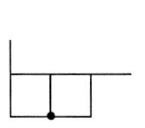 **c** 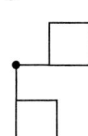 **d**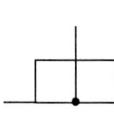

2 Check pupils' rotations are drawn accurately.

3 a B(2, −3)(5, −7)(2, −6) **b** C(−1, 0)(−4, 0)(−5, −3) **c** D(−2, 1)(−5, 5)(−2, 4)

Homework 14.5

1

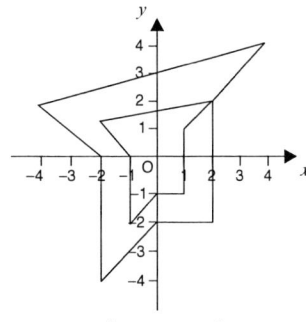

2 a Triangle B coordinates (–4, 1), (–4,–1), (–10, –1)
  b Triangle C coordinates (1.5, 0), (1.5, –0.5), (3, 0)
3 (–9, –3), (–24, –12), (6, –6)

Homework 14.6

1 a Reflection in line $y = x$ **b** Translation through $\begin{pmatrix} 8 \\ 6 \end{pmatrix}$

  c Reflection in line $y = 4$ **d** Enlargement, scale factor 2, centre (4, 5)

  e Rotation, 90° clockwise, centre (–5, 0) **f** Enlargement, scale factor –2, centre (2, –1)

  g Rotation, 90° clockwise, centre (–4, 1) **h** Translation through $\begin{pmatrix} -2 \\ -6 \end{pmatrix}$

  i Rotation, 90° anticlockwise, centre (6, 0)

2 a Enlargement, scale factor pq, centre (0, 0)

  b Translation through $\begin{pmatrix} 6 \\ 0 \end{pmatrix}$ **c** Rotation, 180°, centre (0, 0) **d** Translation through $\begin{pmatrix} a+c \\ b+d \end{pmatrix}$

Homework 15.1

1 Each half of the line should be 3.5 cm.
2 **a** The large angle should be 60°.
  b Both the smaller angles should be 30°.
3 The circumscribed circle will show whether pupils' bisections are accurate. To draw this circle, get the pupils to use a pair of compasses with the point at the intersection of their three lines, and the pencil end on one of the vertices of the triangle; draw the circle. All three vertices should touch the circle (within ±2 mm).
4 In a similar way, the inscribed circle will check their bisections. This time the circle goes *inside* the triangle. The circle should touch the sides of the triangle (not the vertices), again within ±2 mm margin for error.

Homework 15.2

1 and 2 A quick visual check to see pupils have drawn construction lines and a 90° angle.
3 Check that pupils have constructed the triangle accurately.

Homework 15.3

1

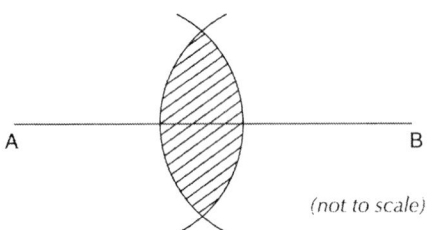

(not to scale)

2 **a**

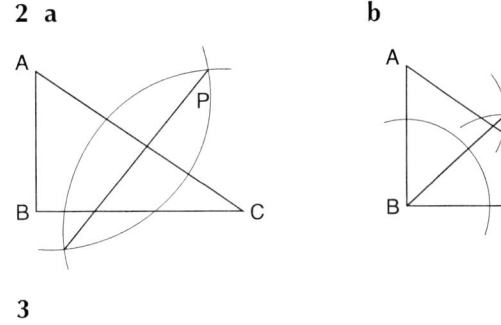

b

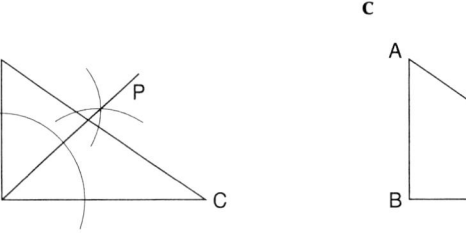

c

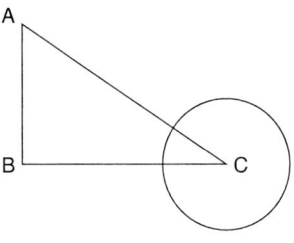

(not to scale)

3

2 cm

(not to scale)

Homework 15.4

1 a

b

c

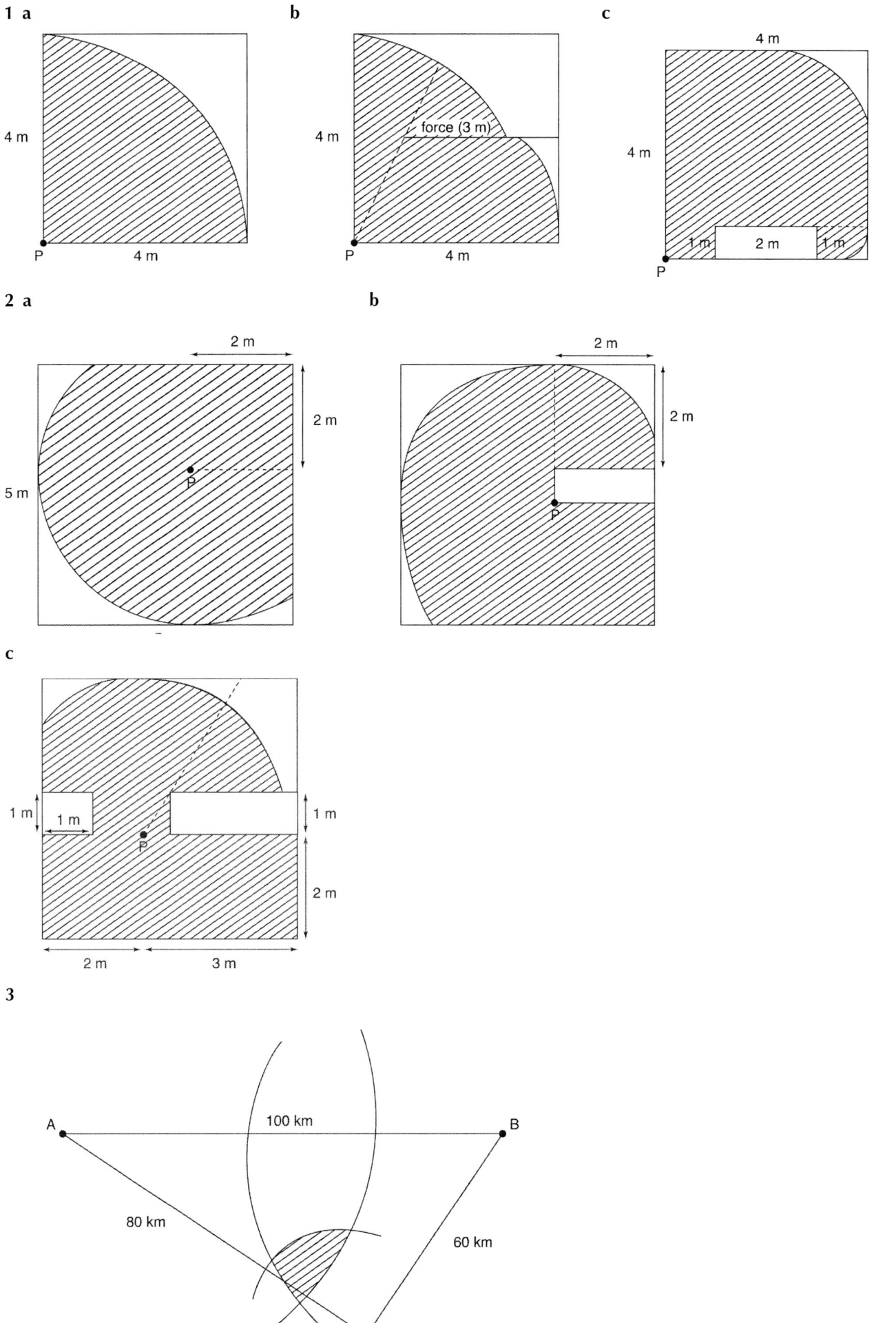

4 m

P

4 m

force (3 m)

4 m

P

4 m

4 m

4 m

P

1 m   2 m   1 m

2 a

b

2 m

2 m

5 m

P

2 m

2 m

P

c

1 m

1 m

1 m

2 m

P

2 m   3 m

3

A

100 km

B

80 km

60 km

C

Chapter 16

Homework 16.1

1 a Yes, scale factor 1.5 **b** No, scale factor of vertical side is 8.8, scale factor of horizontal side is 8.83 so the triangles cannot be similar

2 $x = 0.4\,\text{m},\ y = 12\,\text{m}$

3 2.5 cm   **4** 80 m   **5** $h = 29\,\text{m}$ (nearest metre)

Homework 16.2

1 a 9 : 100   **b** 27 : 1000

2 a 40 500 cm³   **b** 187.5 cm³

3 0.72 m³

4 9.77 hours

5 1.94 kg

Chapter 17

Homework 17.1

1 $2r + \pi r$ **2** $a + b + \sqrt{a^2 + b^2}$ **3** $\dfrac{\pi d}{2} + 2b + d$ **4** $a + 2b + 2c$ **5** $a + b + 2c$

6 $2h + \pi r$ **7** $2a + 2b$ **8** $4a + 4\sqrt{2}a$ **9** $2\pi a$

Homework 17.2

1 $\dfrac{\pi r^2}{2}$ **2** $\dfrac{ab}{2}$ **3** $\dfrac{\pi d^2}{8} + bd$ **4** $ab + \dfrac{c^2}{4}$ **5** $\dfrac{1}{2}(a+b)h$

6 $\dfrac{\pi r^2}{2} + r\sqrt{h^2 - r^2}$ **7** $ac - cd + bd$ **8** $7a^2$ **9** $\dfrac{\pi a^2}{2} + a^2$

Homework 17.3

1 $a^2 b$ **2** $\dfrac{\pi d^2 h}{4}$ **3** $da^2 + bca - dca$ **4** $\dfrac{2}{3}\pi r^3$

5 $\dfrac{2}{3}\pi r^3 + \dfrac{1}{3}\pi r^2 h$ **6** $\dfrac{\pi d^3}{6}$ **7** $h^2(a+b)$ **8** $6a^3$

Homework 17.4

1 a area **b** none of these **c** area **d** none of these **e** length
f length **g** area **h** area **i** volume **j** length
k length **l** length **m** length

2 a 2 **b** 2 **c** 2
d 1 **e** 2 **f** 3

3 b and c

Chapter 18

Homework 18.1

1 a $\overrightarrow{BC}$, $\overrightarrow{CD}$, $\overrightarrow{DE}$, $\overrightarrow{FG}$, $\overrightarrow{GH}$, etc.　　**b** $\overrightarrow{BA}$, $\overrightarrow{CB}$, $\overrightarrow{DC}$, $\overrightarrow{ED}$, $\overrightarrow{GF}$, $\overrightarrow{HG}$, etc.　　**c** 2**a**　　**d** 3**a** + 2**b**

　e 3**a** + 2**b**　　　　　　　　　　　**f** They are equal

　g $\overrightarrow{AY}$　　　　　　　　　　　　　**h–m** See grid below　　　　　　　**n** $\overrightarrow{EQ}$, $\overrightarrow{JV}$, $\overrightarrow{DP}$, $\overrightarrow{IU}$

　o $\overrightarrow{SA}$, $\overrightarrow{XF}$, $\overrightarrow{TB}$, $\overrightarrow{YG}$　　　　　　**p** any vector equivalent to **a** + **b** such as $\overrightarrow{AG}$, $\overrightarrow{BH}$, $\overrightarrow{CI}$, etc.

Answers to parts h–m

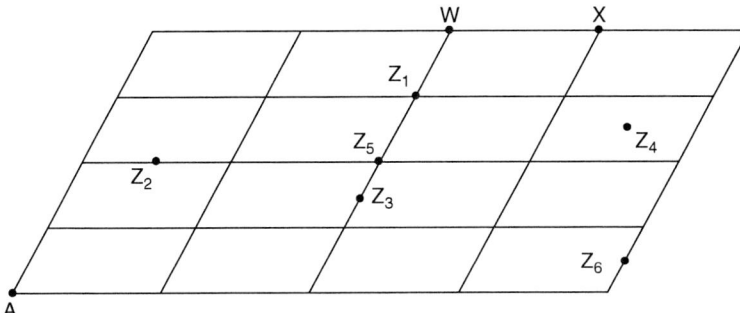

Homework 18.2

1 a　i **b** − **a**　　　　**ii** $\frac{1}{2}$(**b** − **a**)

　b Going from O to M gives the same result as going from O to A then A to M

　c　i **a** + $\frac{1}{2}$(**b** − **a**) or $\frac{1}{2}$**a** + $\frac{1}{2}$**b** or $\frac{1}{2}$(**a** + **b**)　　**ii** $\frac{1}{2}$(**a** − **b**)

2 a $\overrightarrow{EF}$ = $\overrightarrow{EG}$ + $\overrightarrow{GF}$ (see diagram below), so

　　$\overrightarrow{EF}$ = −**a** + **b** = **b** − **a**

　b　i −**a**　　**ii** −**b**　　**iii** **a** − **b**

　c 0, starts at A and ends at A, so overall no movement

　d　i 2**a** − 2**b**　　**ii** 2**a**　　**iii** 2**a** − **b**

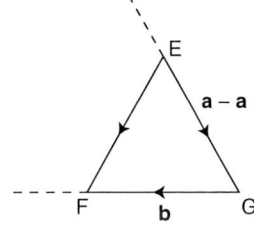

Chapter 19

Homework 19.1

19.1a

1 15	**2** 19	**3** 9	**4** 50	**5** 2
6 2	**7** 25	**8** 75	**9** $\frac{1}{25}$ or 0.04	**10** $30\frac{2}{3}$
11 13	**12** 16	**13** 5	**14** 29	**15** 21
16 11	**17** 91	**18** 36		

19.1b

1 $6 + 3a$	**2** $20 - 8b$	**3** $c^2 + 2c$	**4** $2d^2 + 16d$	**5** $24e - 8e^2$
6 $10f^2 + 15fy$	**7** $g^3 - 8g$	**8** $h^4 + 8h$	**9** $4i^3 - 12i$	**10** $10j^3 - 10j$
11 $11k^4 + 33k$	**12** $25l^3 + 5l^2$	**13** $6m^4 + 8m^3$	**14** $15n^3 - 10nx$	
15 $20p^4 - 15p\text{w}$				

19.1c

1 a $9a$ **b** $7e$ **c** $9p^2$ **d** $9a^2b$

2 a $22 + 5a$ **b** $11 + 5b$ **c** $2 + 2c$ **d** $3d + 25$

  e $4e + 2ex + 3x$ **f** $12f + 13fy + 24y$ **g** $g^2 + 9g$ **h** $14h^2 - 3hw$

  i $18iv + 6iu - 8uv$ **j** $8j^3 - 4j^2$

Homework 19.2

1 $6(a + 3b)$ **2** $5(2c + d)$ **3** $2(5e - 3f)$ **4** $2(3g - 4h)$ **5** $3a(b + c)$

6 $4d(2e + f)$ **7** $g(5g + 2)$ **8** $6h(h - i)$ **9** $3k(2j - 3a)$ **10** $6lm(l - 2)$

11 $20q(4pr + 3st)$ **12** $2(5a^2 + 3a + 2)$ **13** $2e(2d + 4f + g)$ **14** $2jk(2k + 1 - 4j)$

15 The terms have no factors in common.

Homework 19.3

19.3a

1 $a = 8$ **2** $b = -5$ **3** $c = 9$
4 $d = 7$ **5** $e = 3$ **6** $f = 1.5$

19.3b

1 9 **2** 12 **3** 15 **4** $2\frac{1}{5}$ (or 2.2) **5** -1
6 -3 **7** $-\frac{3}{2}$ (or -1.5) **8** $-\frac{7}{2}$ (or-3.5)

19.3c

1 4 **2** 2 **3** 5 **4** 7 **5** 3
6 10 **7** 6 **8** 4

19.3d

The pupils may use different letters to those given in the answer.
1 $P = rh + b$, £130 **2** $P = rh + b$, £137.50
3 $C = 3d$, 180 cm **4** $C = pn$, £4.25
5 $22h + 50 = 204$, 7 hours
6 $m + 2m + (m + 14) = 82$. Fred has £17, Bill has £34, Jack has £31

Homework 19.4

1 5.1 **2** 2.5 **3** 3.4 **4** 4.6 m

Homework 19.5

19.5a

1 a $x = 2, y = 4$ **b** $x = 4, y = 1$ **2 a** $x = 4, y = 2$ **b** $x = \frac{1}{2}, y = 4$

19.5b

1 a $2x + y = 30.97$ $3x + 2y = 52.95$ **b** $x = £8.99$ $y = £12.99$ **c** cost $= £57.94$

2 Long $= 600$ mm, short $= 400$ mm, total $= 5400$ mm

Homework 19.6

1 $x = \dfrac{A}{3}$

2 $x = \dfrac{B+5}{3}$

3 $x = \dfrac{C}{2\pi}$

4 $x = \dfrac{D-2d}{2}$ or $\dfrac{D}{2} - d$

5 $x = \sqrt{E+22}$

6 $x = \dfrac{F+G}{4}$

7 $x = L^2 - H$

8 $x = \sqrt{\dfrac{J}{5}}$

9 $x = \sqrt{\dfrac{K-L}{8}}$

Chapter 20

Homework 20.1

1 $x^2 + 5x + 6$
5 $y^2 - 4$
9 $x^2 - 6x$

2 $x^2 + 11x + 30$
6 $s^2 - 25$
10 $x^2 + 10x$

3 $2x^2 + 16x + 30$
7 $c^2 + 4c + 4$

4 $14 + 9x + x^2$
8 $25x^2 - 20x + 4$

Homework 20.2

1 $(x + 1)(x + 3)$
4 $(x - 4)(x + 4)$

2 $(x - 7)(x + 5)$
5 $(2x - 9)(2x + 9)$

3 $(x + 9)(x - 2)$
6 $(x - y)(x + y)$

7 $(x - 2b)(x + 2b)$

8 $(x - \dfrac{1}{2})(x + \dfrac{1}{2})$

9 $(2x + 3)(x + 1)$

10 $(3x + 4)(x - 7)$

11 $(4x - 3)(x - 5)$

12 $(3x + 2)(4x + 5)$

Homework 20.3

1 $x = \dfrac{1}{3}$ or -3

2 $x = -1$ or 6

3 $x = 4$ or 2

4 $x = \dfrac{1}{2}$ or -1

5 $x = \pm \dfrac{3}{2}$

6 $x = \dfrac{3}{2}$ or $\dfrac{1}{2}$

Homework 20.4

1 $x = \dfrac{1}{3}, -1$

2 $x = \dfrac{-2 \pm 2\sqrt{7}}{6} = \dfrac{-1 \pm \sqrt{7}}{3}$

3 $x = 2.62, 0.38$

4 $x = -0.64, -0.86$

Homework 20.5

1 a $(x + 10)^2 - 100$

b $(x - \dfrac{19}{2})^2 - \dfrac{361}{4}$ [or $(x - \dfrac{19}{2})^2 - 90.25$]

c $(x - 9)^2 - 86$

d $(x + \dfrac{17}{2})^2 - \dfrac{269}{4}$ [or $(x + \dfrac{17}{2})^2 - 67.25$]

2 $x = \pm\sqrt{69} - 8$
3 $x = 14.66$ or 0.34

Homework 20.6

1 a -23, no solution

b 60, two solutions

c 28, two solutions

2 $\dfrac{1 \pm \sqrt{7}}{3}$

3 2 and 5

Chapter 21

Homework 21.1

1 20 km
2 40 km/h
3 10 km
4 30 km/h
5 120 km/h
6 90 km/h
7 Trish by 10 mins
8 66 km/h
9 55 km/hr
10 0.6̇ km/min or 40 km/h

Homework 21.2

1 **a** 2 m/s^2 **b** 6 m/s **c** 3 m/s^2

2
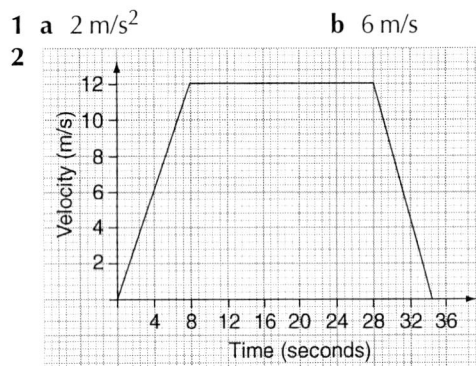

Homework 21.3

1
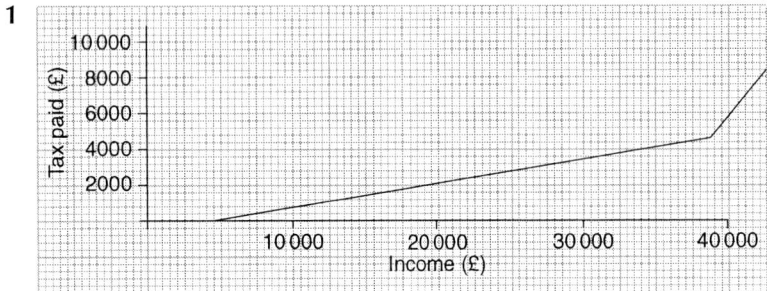

£500 (actually £502)

2
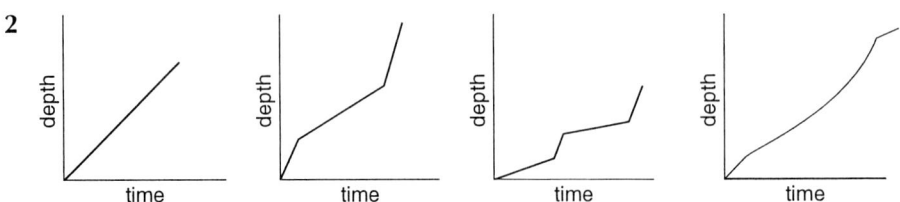

Homework 22.1

1 14.2 cm **2** 31.0°, 31.0° and 118°
3 AD = 12.6 cm, $x = 28.0°$
4 20.2 cm
5 17.6 cm

Homework 22.2

1 8.66 cm
2 a 28.3 cm **b** 24.4 cm **c** 24.4 cm
3 a 13 cm **b** 16.4 cm **c** 17.7°
4 a 34.5° **b** 45.0°

Homework 22.3

1 a 17°, 163° **b** 41°, 319° **c** 74°, 254°
   d 221°, 319° **e** 145°, 215°
2 30°, 330°
3 20°, 140°
4 51°, 269°

Homework 22.4

22.4a
1 a 14.4 cm **b** 9.79 cm **c** 14.1 cm
2 a 35.1° **b** 35.9° **c** 21.8°
3 38.3°, 141.7°
4 a 32.8 km **b** 45.7 km

22.4b
1 a 3.72 cm **b** 3.62 cm **c** 3.50 cm
2 a 38.1° **b** 63.8° **c** 143°
3 33.6°, 50.7°, 95.7°
4 2.69 km

Homework 22.5

1 $\dfrac{4}{\sqrt{41}}$ **2** $\dfrac{\sqrt{22}}{5}$ **3** $\dfrac{\sqrt{5}}{2}$ **4** $\dfrac{1}{\sqrt{2}}$ **5** $\dfrac{\sqrt{7}}{5}$ **6** $\dfrac{\sqrt{5}}{10}$ **7** $3\sqrt{3}$ cm^2

Homework 22.6

1 a 4.52 cm^2 **b** 14.0 cm^2 **c** 5.09 cm^2
2 a 5.37 cm^2 **b** 5.60 cm^2 **c** 4.32 cm^2
3 7.20 cm **4** 64.3 cm^2

Homework 23.1

1, 3 and 4

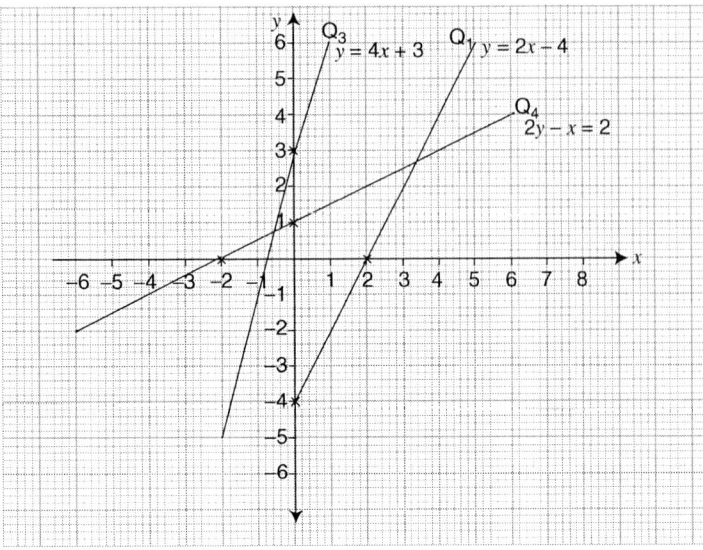

2 gradient $= \dfrac{1}{3}$

Homework 23.2

1 a $y = x + 2$ **b** $y = 3x$ **c** $y = 2x - 2$

2 a $y = -x + 2$ **b** $y = -2x$ **c** $y = -x - 2$

3 a $y = \dfrac{1}{2}x + 1$ **b** $y = -\dfrac{1}{2}x + 3$ **c** $y = \dfrac{1}{3}x - 3$

Homework 23.3

1 a $-17\ °C$ or $-18\ °C$ (accurate answer is $-17.\dot{7}\ °C$)

    b Any good approximation to $\dfrac{11}{20}$ or 0.55 (accurate answer is $0.\dot{5}$ or $\dfrac{5}{9}$)

    c $m = 0.55$, $c = -18$, rule is $C = 0.55F - 18$ (or any correct interpretation of their own results)

2 a 1.1   **b** £6   **c** $\text{Cost} = 1.1M + 6$

3 $(2, 4)$

Homework 23.4

1 $\dfrac{1}{2}$ **2** $y = x + 4$ **3** $y = -\dfrac{1}{4}x + 2$

4 $y = -\dfrac{1}{2}x + 2$ **5** $y = \dfrac{1}{4}x + 1$ **6** $y = -2x - 3$

Homework 24.1

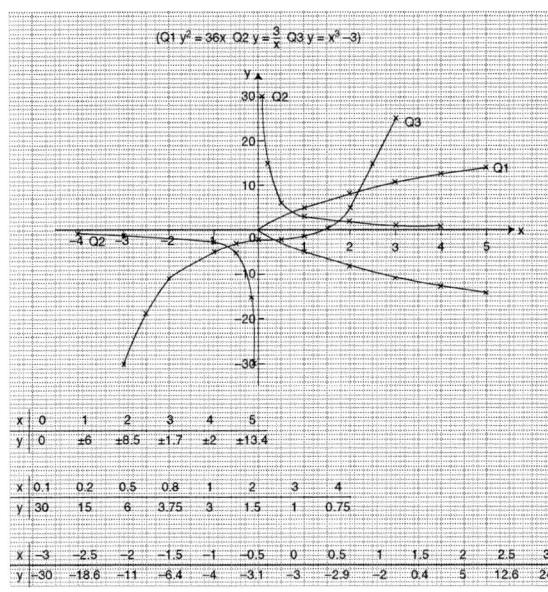

(Q1 $y^2 = 36x$ Q2 $y = \frac{3}{x}$ Q3 $y = x^3 - 3$)

x	0	1	2	3	4	5
y	0	±6	±8.5	±1.7	±2	±13.4

x	0.1	0.2	0.5	0.8	1	2	3	4
y	30	15	6	3.75	3	1.5	1	0.75

x	−3	−2.5	−2	−1.5	−1	−0.5	0	0.5	1	1.5	2	2.5	3
y	−30	−18.6	−11	−6.4	−4	−3.1	−3	−2.9	−2	0.4	5	12.6	24

1 a $\sqrt{3}$ = 1.7 and −1.7, $\sqrt{4}$ = 2 and −2, $6\sqrt{3}$ = 10.4 and −10.4, $6\sqrt{4}$ = 12 and −12

   b y = 9.5 and −9.5   **c** x = 0.4

2 a 15, 6, 1   **b** y = 1.2   **3** −30, −11, −4, −2.9, 0.4, 12.6

Homework 24.2

1 a   27 and 0.11   **b**   **c** y = 15.6

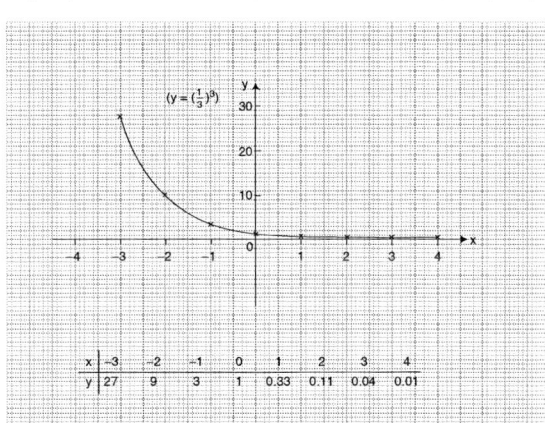

$(y = (\frac{1}{3})^x)$

x	−3	−2	−1	0	1	2	3	4
y	27	9	3	1	0.33	0.11	0.04	0.01

2 a   0.04 and 9   **b**   **c** x = 3.1

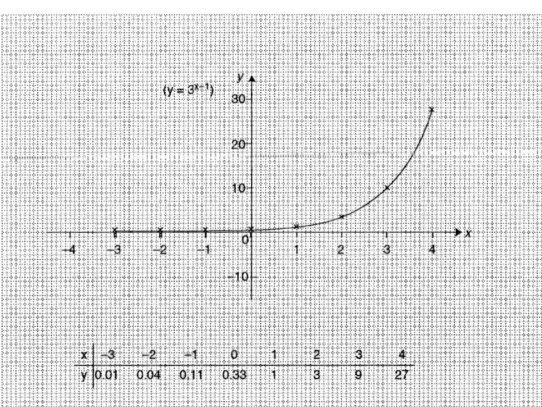

$(y = 3^{x-1})$

x	−3	−2	−1	0	1	2	3	4
y	0.01	0.04	0.11	0.33	1	3	9	27

Homework 24.3

1 a Correct graph of $y = \sin x$ **b** 120°   **c** 240°
2 a Correct graph of $y = \cos x$ **b** 330°   **c** 150°, 210°

Chapter 25

Homework 25.1

1 a $\dfrac{7a}{12}$ **b** $\dfrac{8a-3}{4}$ **c** $\dfrac{a}{4}$ **d** $\dfrac{a^2}{12}$ **e** $\dfrac{1}{2a}$

2 $a = 5$ **3 a** $\dfrac{4}{3}$ or $1^1/_3$ **b** 1

4 $\dfrac{2(5a-4)}{(a-2)(a+1)}$ **5** $a = \dfrac{1}{2}$ or $a = 6$

Homework 25.2

1 (4, 5) (−5, −4)   **2** (−5, −5) (7, −1)   **3** (3, 10) (4, 18)   **4** (6, 21) (−1, 0)

Homework 25.3

1 a 12, 14   **b** 32, 64   **c** 21, 34   **d** 15, 21
2 a $2n - 1$ **b** $10n + 90$ **c** $5n - 3$ **d** $5n + 7$
3 a $10n - 4$, 496   **b** $n + 7$, 57   **c** $8n - 7$, 393   **d** $10n + 80$, 580
4 a $4n - 1$, 99   **b** $3n - 10$, 101   **c** $5n + 12$, 102   **d** $2n + 18$, 100

Homework 25.4

1 a i 22   **ii** $5n + 2$ **iii** 102   **b** 40
2 a i 12   **ii** 202   **b** 8580 m

Homework 25.5

1 a 44, 57   **b** $n^2 + 8$ **c** 2508
2 a $2n^2 - 3n + 10$ **b** 4860

Homework 25.6

1 $a = \dfrac{11}{3}$ **2** $b = \dfrac{y-x}{2a}$ **3** $c = \dfrac{px-ty}{t-p}$ or $\dfrac{ty-px}{p-t}$ **4** $h = \dfrac{V - \frac{4}{3}\pi r^3}{\pi r^2}$

5 $e = \dfrac{pq+5}{2+p}$ **6** $e = \dfrac{pq+5}{2+p}$

Homework 26.1

1 a $a < 30$ **b** $b \geq 19$ **c** $c > -8$ **d** $d \leq 5$

2 a 9   **b** 1   **c** 2   **d** 10

3 a $-1 \leq x \leq 4$ **b** $1.5 \leq x < 6.5$

4 $-10 \leq x \leq 10$

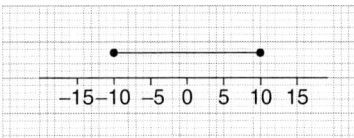

Homework 26.2

1

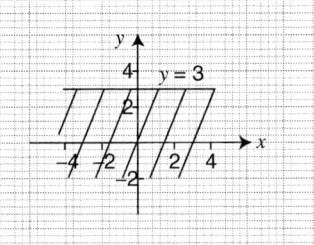

2

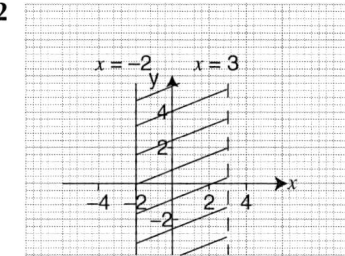

3

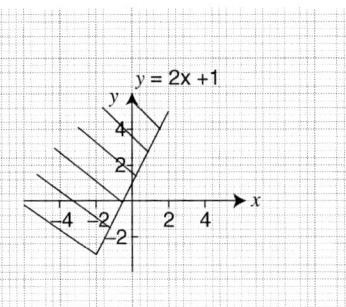

c No

4

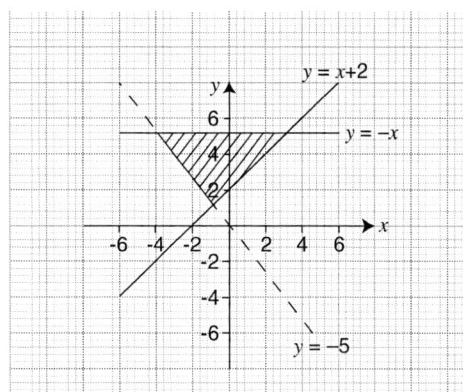

a Yes
b Yes
c No

Homework 26.3

1 a may be true, for example, $e = 10$ **b** must be false as $e + f$ can't be < 20 and ≥ 21 at the same time
  c may be true, for example, $e = 5, f = 5$ **d** may be true, for example, $e = 15, f = 1$

2 a $50x + 70y \leq 500$ or $5x + 7y \leq 50$ **b** $x \geq y + 3$ or $y \leq x - 3$

3 a **i** number of seats required is $15x + 50y \geq 250$ which cancels down to $3x + 10y \geq 50$
    ii number of 50-seat coaches ≤ 4 **iii** number of 15-seat coaches ≤ 10
  b **i** Yes **ii** Yes **iii** No **iv** Yes **v** Yes
  c v

Homework 27.1

1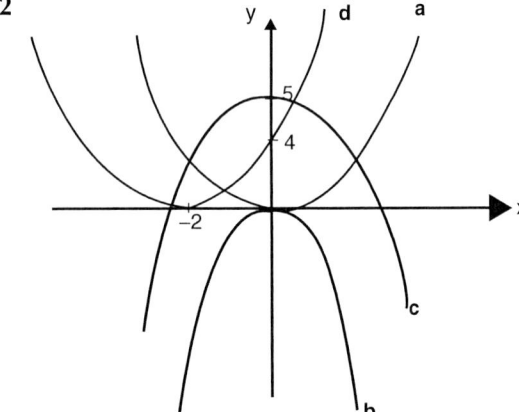

$y=x^2, y=2x^2, y=5x^2, y=\frac{1}{3}x^2$

  e all are a stretch in the y-direction
  b has a scale factor 2
  c has a scale factor 5
  d has a scale factor $\frac{1}{3}$

2

  e b is a reflection in the x-axis, then a stretch in the y-direction of scale factor 2

  c is a reflection in the x-axis then a translation $\begin{pmatrix} 0 \\ 5 \end{pmatrix}$

  d is a translation $\begin{pmatrix} -2 \\ 0 \end{pmatrix}$

3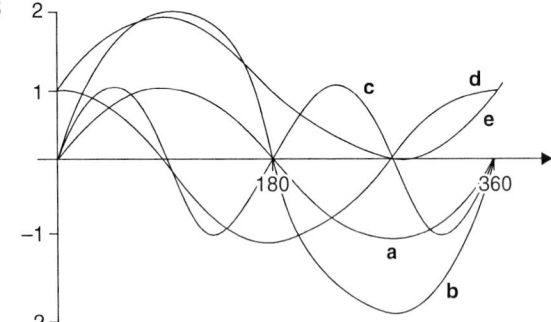

Homework 28.1

The solutions can be found in the Pupil Book.

Homework 28.2

1 a Mean $= \dfrac{1+2+3+4+5}{5} = \dfrac{15}{5} = 3$ $\qquad\qquad$ $\therefore$ square of mean $= 9$

Mean of squares $= \dfrac{1+4+9+16+25}{5} = \dfrac{55}{5} = 11$ $\qquad$ $\therefore$ difference $= 11 - 9 = 2$

b Square of the mean $= \left(\dfrac{n+(n+1)+(n+2)+(n+3)+(n+4)}{5}\right)^2 = \left(\dfrac{5n+10}{5}\right)^2$

$\qquad\qquad\qquad = (n+2)^2 = n^2 + 4n + 4$

Mean of the squares $= \dfrac{n^2+(n+1)^2+(n+2)^2+(n+3)^2+(n+4)^2}{5}$

$\qquad\qquad = \dfrac{n^2+\left(n^2+2n+1\right)+\left(n^2+4n+4\right)+\left(n^2+6n+9\right)+\left(n^2+8n+16\right)}{5}$

$\qquad\qquad = \dfrac{5n^2+20n+30}{5} = n^2 + 4n + 6$

So difference between them $= (n^2 + 4n + 6) - (n^2 + 4n + 4) = 2$

Chapter 1

1 43.7 matches

2 a $5 < t \leqslant 10$ **b**

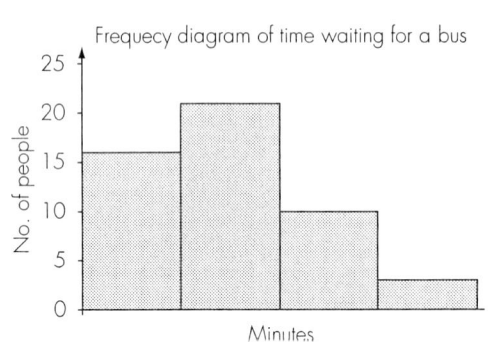

Frequecy diagram of time waiting for a bus

(y-axis: No. of people, x-axis: Minutes)

3 a Level 6   **b** Level 5   **c** Level 3.2

d All results are all or below diagonal which shows equal grades in both subjects. Therefore all students as good or better at French. Only 5 are equal to both subjects, so it is reasonable to say the students are better at French.

4 22.3 minutes

5 a 140 km   **b** 100 km/h

6 a The statement contains two facts, but you can only give one reply; there is no space to reply "Don't know"; the wording implies Jack is expecting a negative response.

b There is a full range of possible answers; the question is neutral.

7 a

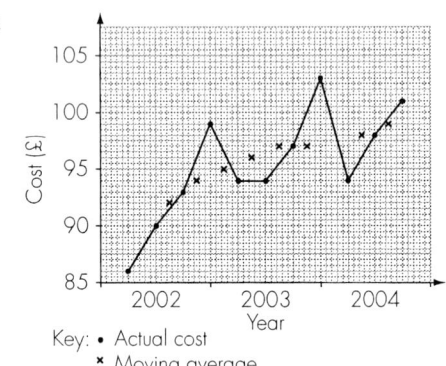

Key: • Actual cost
     × Moving average

(y-axis: Cost (£), x-axis: Year)

8 a stratified sampling

b 1 large-load vehicle, 13 light vans, 2 cars

9 a

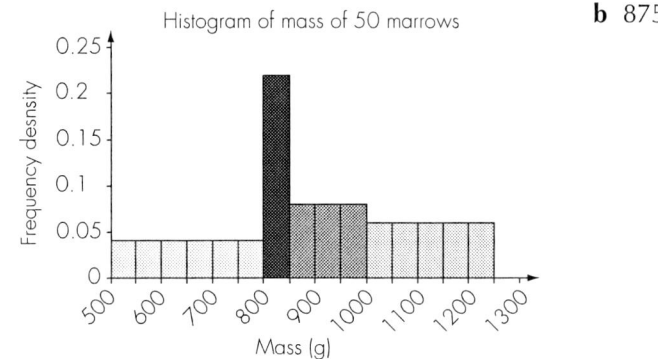

Histogram of mass of 50 marrows

(y-axis: Frequency desnity, x-axis: Mass (g))

b 875 g

b Draw a trend lane throughout the *x*s. It will cross the last vertical line of the graph paper at about £100. Use the value as the answer to **b**.

10 a 102   **b** 85 to 120, IQR = 35

11 a i 40 marks   **ii** 50–30, IQR = 20 marks

b i 25 marks

ii Any histogram with 20 squares altogether, the 30–40 column with 3 squares (frequency density = 1.5), and the 40–50 column with 5 squares (frequency density = 2.5).

Chapter 2

1 22.3 minutes

2 a i Brian's median score = 100   **ii** Brian's IQR is about 13

b i George - smaller range and smaller IQR   **ii** Brian – lower median

3 The girls had a larger range and a larger inter-quartile range. The median mark for the girls was lower than the median for the boys.

4 Limits are LQ and UQ; 11 minutes to 19 minutes

1 **a** 0.28   **b** 1600 pupils

2 **a i** 047   **ii** 5 red, 3 green, 2 blue
   b He has taken a very small sample so his results are not reliable but even if he took a much larger sample he could still fail to pick a blue ball. He cannot be sure there are no blue balls.

3 **a**

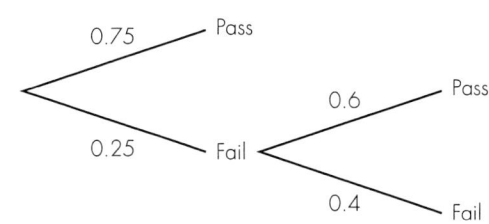

   b 0.52

4 **a** 0.3   **b** 1
   c

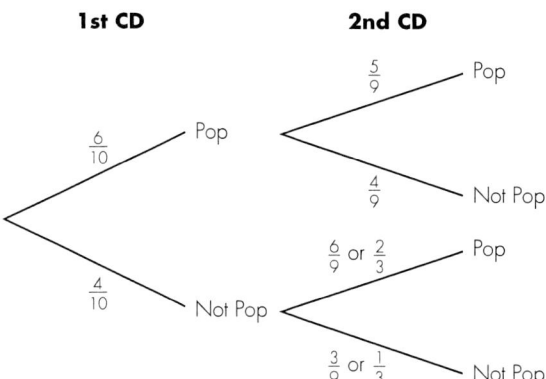

   d $\frac{1}{21}$

   ii 0.1   **b** 0.729

5 **a i**

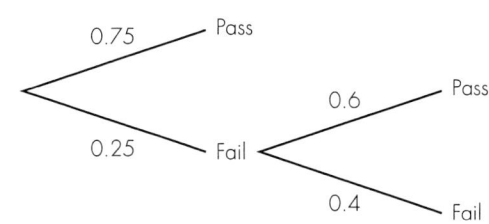

6 $\frac{13}{28}$

7 **a** $\frac{1}{56}$ **b** $\frac{15}{28}$ **c** $\frac{55}{56}$ **d** $\frac{1}{21}$

1 6 weeks

2 14 boxes

3 **a** 90   **b** 240   **c** 6

4 **a** $2 \times 2 \times 7$ **b** 84

5 Approximation $= \dfrac{30}{4 \times \frac{1}{2}} = 15$ Gemma is correct.

6 4200

7 40 beats

8 **a** $x = 5$
   b $3 \times 2 \times 5 \times 5$

9 **a** $p = 2, q = 5$
   b 10

10 **a** $a = 2, b = 5$ (or vice versa)   **b** There are two possible solutions, 1 and 6 or 2 and 3.

Chapter 5

1 £332.80

2 a 24.2 sec   b i 22.5 sec   ii Darren   iii Chris

3 £141

4 Pops: £19.20, Sounds: £20.16, Pops offers better value

5 £220   6 $\frac{9}{40}$ 7 $4\frac{1}{12}$ pints   8 5 tins

9 a $3\frac{1}{8}$, $\frac{22}{7}$, $\frac{256}{81}$, $\sqrt{10}$ b $\frac{22}{7}$

10 a 44%   b 216 kangaroos

11 8% decrease

12 Estimate: 80% (78.4%)

13 32% increase   14 £7375.53   15 $\frac{5}{6}$

16 a £3025   b £1200

17 a Mr Dale's   b £76.50

Chapter 6

1 4 minutes

2 a 8 glasses   b 50 ml

3 £5800 : £29 000

4 12 black beads

5 a £105   b 70%

6 40°

7 141 adults

8 a 140 km   b 100 km/h

Chapter 7

1 a i d^5 ii e^{-7} iii $6g^4/h^5$

2 5.9×10^9 km

3 a 1.75×10^6 b 8.2×10^{-3}

   c 0.049   d 2.6×10^6

4 a $3ab^2$ b $81x^{12}y^8$

5 $\frac{8}{15}$

6 $a = 5, b = 2$

7 $x = 5\sqrt{2}$

8 b $\frac{41}{110}$

9 a ii $14 + 4\sqrt{6}$

10 $\dfrac{2\sqrt{3} + 3}{3}$ or $\dfrac{2\sqrt{3}}{3} + 1$

Chapter 8

1 a

x	–2	–1	0	1	2	3
y	15	5	–1	–3	–1	5

b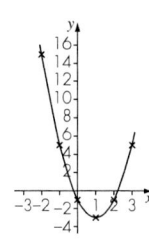

c i The solutions are where the graph cuts the x-axis.
   ii –0.22

2 a

x	–3	–2	–1	0	1	2	3
y	0	4	6	6	4	0	–6

b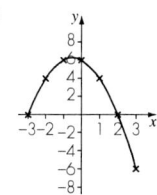

c $x = 1.56$ or $x = –2.56$

3 a $(0.5, –12.25)$ **b** $x^2 – 2x – 8 = 0$
4 a $x = 0.8$ or $x = –3.8$ **b** $x = 0.4$ or $x = –2.4$

Chapter 9

1

x	25	100	400
y	10	20	40

2 a $E = 4000\,v$ **b** 3.6 m/s

3 a $y = 4x^{-\frac{1}{3}}$ or $y = \dfrac{4}{\sqrt[3]{x}}$ **b i** 20   **ii** 8

4 19.4 cm

5 128

6 $\frac{3}{4}$

7 b $h \propto t^2$

8 a 2.5   **b** 0.25   **c** 250   **d** 50, –50

9 a 10   **b** 3.375

10 a 48π **b** 9

11 27 Hz

12 a 640   **b** 4

13 40

Chapter 10

1 a 1845   **b** 1854
2 215 miles
3 a 79.5 cm   **b** 40.5 cm   **c** 3260.25 cm^2
4 18.95 cm and 21.16 cm
5 1.36
6 5.63 cm
7 2.92
8 50 crates
9 7.39 m/s
10 No
11 a 5.197 m   **b** Yes

Chapter 11

1 **a** $a + b = 10$, with a or $b \neq 5$
2 23.1 cm
3 **a** 12π cm^2 **b** $(4\pi + 12)$ cm
4 85π cm^2
5 No, volume of water in cylinder is 1385 cm^3
6 4.96 g/cm^3 or 5.0 g/cm^3
7 634 cm^3

Chapter 12

1 177 m
2 3.23 m
3 **a** 11.2 cm   **b** 37.5 cm
4 **a** 4.33 m   **b** 78.5°
5 **a** 35.0°   **b** 17.9 cm
6 4.16 m
7 **a** 10 cm   **b** $\frac{4}{5}$

Chapter 13

1 45°
2 20
3 **a** 33°   **b** 123°
4 **a i** 140°   **ii** 70°   **b i** 124°   **ii** 42°
5 **a i** 100°   **ii** 130
   b i cyclic quadrilateral, so $12x = 180°$ **ii** 120°
6 $x = 36°$

Chapter 14

1 **a** C and D

2 **a**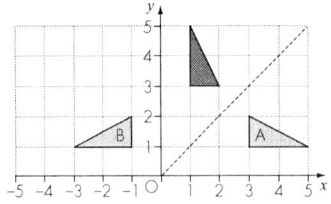
   b reflection in $x = 1$

3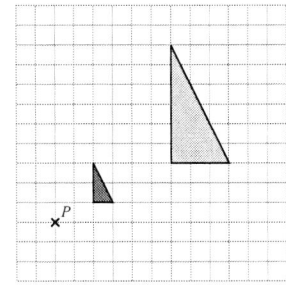

4 **a** A rotation of 90° clockwise about (0, 2)   **b**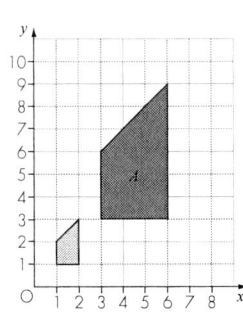

5 **a** $\frac{1}{2}$ **b** $(-2, -1)$

6 a A rotation of 180° about the point (1, 3)   **b**

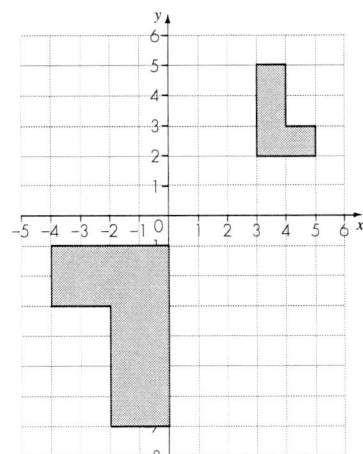

Chapter 15

1

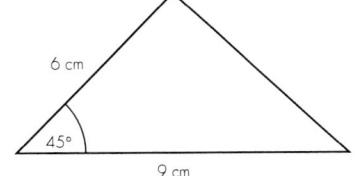

2

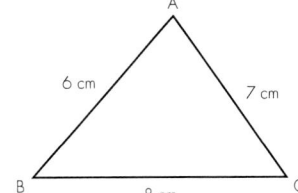

3 a

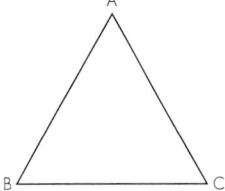

b BC = 9.2 cm

4

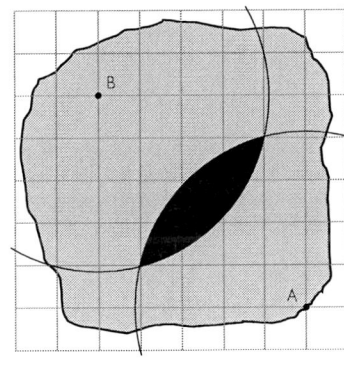

5 a

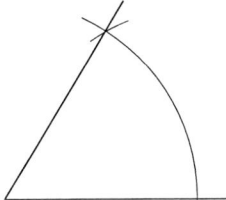

b

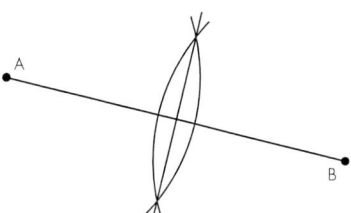

6

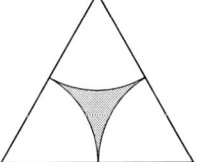

7 a

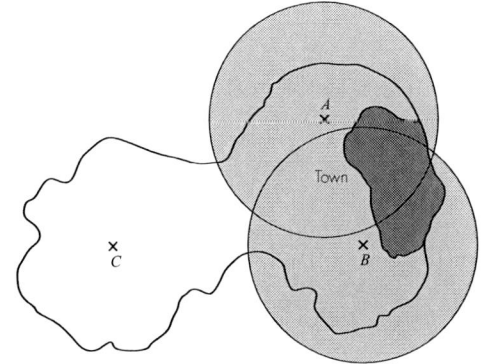

b 50 km

8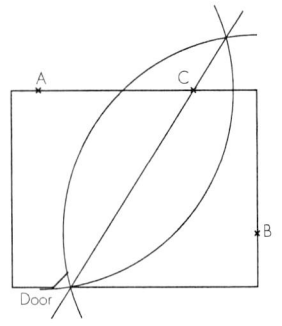

9

10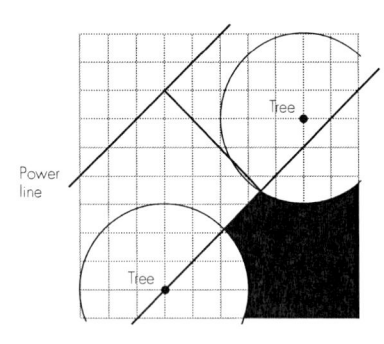

11

---

1   No. The ratios of the sides are not the same
2   CD = 2.7 cm
3   PB = 20 cm
4   RY = 20 cm
5   £90
6   **a** Volume increases by cube of length, so if length is multiplied by 2, volume is multiplied by 2^3 or 8.
    b The volume has increased by a factor of 5.95. So the claim is not justified.

Chapter 17

1   **a** V **b** L
2   **a** V **b** L **c** A
3   **a** Inconsistent, should be r^3 **b** Inconsistent, 2-D + 1-D
4   **a** V **b** A **c** N
5   **a** Formula 2   **b** Consistent for area
6   **a** A **b** N **c** V

Chapter 18

1   $\frac{1}{4}\mathbf{s} + \frac{3}{4}\mathbf{t}$
2   **a i** $\mathbf{a}+\mathbf{b}$ **ii** $2\mathbf{a}+\mathbf{b}$ **iii** $\mathbf{b}-\mathbf{a}$ **b** $\mathbf{a}+2\mathbf{b}$ **c** $3\overrightarrow{OD} = \overrightarrow{OF}$, so O, D and F lie on a straight line
3   **a i** $\frac{1}{2}\mathbf{a}-\frac{1}{2}\mathbf{c}$ **ii** $\frac{1}{2}\mathbf{b}-\frac{1}{2}\mathbf{c}$ **iii** $\frac{1}{2}\mathbf{a}-\frac{1}{2}\mathbf{b}$ **b** a parallelogram, since $\overrightarrow{PS} = \overrightarrow{QR}$ and $\overrightarrow{PQ} = \overrightarrow{SR}$
4   **a** $4\mathbf{b}-4\mathbf{a}$ **b** $2\mathbf{b}-\mathbf{a}$ **c** $2\overrightarrow{PQ} = \overrightarrow{PR}$
5   **a i** $\mathbf{a}-2\mathbf{b}$ **ii** $\frac{1}{3}\mathbf{a}+\frac{1}{3}\mathbf{b}$ **iii** $\frac{2}{3}\mathbf{a}+\frac{2}{3}\mathbf{b}$ **b** a trapezium, since $\overrightarrow{OP} = 2\overrightarrow{MQ}$
6   **a i** $\mathbf{a}+\mathbf{c}$ **ii** $2\mathbf{a}$ **b** a trapezium, since $\overrightarrow{OA}$ is parallel to $\overrightarrow{CB}$
    c i $\frac{2}{3}\mathbf{a}+\frac{1}{3}\mathbf{c}$ **ii** P lies on OB with $OP = \frac{1}{3}OB$
7   **i** $\mathbf{b}-\mathbf{a}$ **ii** $\frac{1}{2}\mathbf{b}-\frac{1}{2}\mathbf{a}$ **iii** $\frac{1}{2}\mathbf{a}+\frac{1}{2}\mathbf{b}$ **iv** $\frac{1}{3}\mathbf{a}+\frac{1}{3}\mathbf{b}$ **v** $\frac{1}{3}\mathbf{b}-\frac{2}{3}\mathbf{a}$ **vi** $-\mathbf{a}+\frac{1}{2}\mathbf{b}$ **b** $k = \frac{3}{2}$

Chapter 19

1 a $12x - 15$ **b** $x = 3\frac{1}{2}$

2 $x = 1\frac{3}{4}$

3 $x = 5.3$ (1 decimal place)

4 a $14x + 1$ **b** $2x^3 - 6x^2$ **c** $x^2 + x - 2$

5 a $t = \dfrac{2 - s}{3}$ **b** $x = 36$

6 a $x = 3\frac{1}{2}$ **b** $y = 2\frac{1}{5}$ **c** $z = 1\frac{1}{2}$

7 $x = 4.8$ (1 decimal place)

8 a i $s^3 + 6s$ **ii** $7x - 2$ **iii** $n^2 + 6n + 9$ **b i** $(a + 0)(2a + 1)$ **ii** $4xy^2(2x^2 - y)$

9 a $11x + 14$ **b** $4x^2 - 2x^3$

10 $t = \sqrt{10 - s}$

11 $x = 2y - 1$

12 $x = \frac{1}{2}, y = 4$

13 $x + y = 25$, $4x + 3y = 84$, 1 hour

14 a $(x + 1)^2 = x + 6$ **b** 1.8 (1 decimal place)

Chapter 20

1 a $2(4p - 3)$ **b** $r(r + 6)$ **c** s^6 **2** 6

3 a $x^2(x^3 - 4)$ **b i** $(x + 2)(x - 5)$ **ii** $x = 5$ or $x = -2$ **4** $6p^2 - 13pq - 5q^2$

5 a $a^2 - b^2$ **b i** $(x - 2)(x - 18)$ **ii** $x = 2$ or $x = 18$

6 a width $= 12\frac{1}{2} - x$ or width $= \dfrac{25 - 2x}{2}$

b $x(12\frac{1}{2} - x) = 38$
$12\frac{1}{2}x - x^2 = 38$
$25x - 2x^2 = 76$
$0 = 76 + 2x^2 - 25x$
$2x^2 - 25x + 76 = 0$

c $x = 7.28$ or $x = 5.22$

7 a $a = 5, b = 15$ **b** 15   **8 a** $3(x - 2y)(x + 2y)$

9 a $(2n + 3)(n + 1)$ **b** 23×11 **10** $x = 1.74$ or $x = -5.74$

11 a $a = 3, b = -12$ **b** $-3 \pm 2\sqrt{3}$ **12 a** $(x - 2)(x - 6)$ **b** $y = 1$ or $y = 5$

Chapter 21

1 a

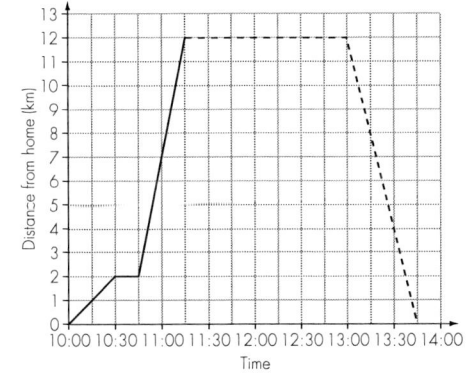

b 16 km/h

2 a He has stopped, perhaps he's having his lunch.

b 16 miles

c 8 miles/h

d

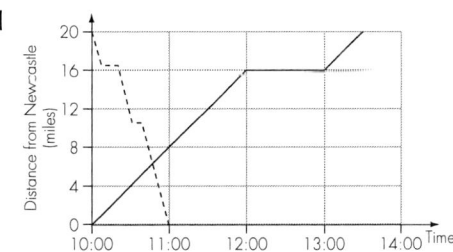

3 68 miles/h   **4** 20 km/l

5 a Graph D. The height of the liquid will rise streadily, then rise increasingly faster, and finally rise steadily again, but fastest of all.

b Any container with vertical sides.

6 a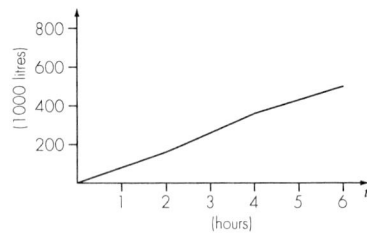

b 50 l/h

7 a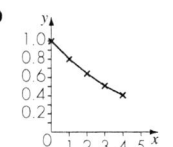

b 25 km/h

Chapter 22

1 a 16.7 cm   **b** 24.6°   **c** 22.6°   **2** BC = 10 cm   **3** 5.52 cm

4 94.9°   **5** 14.2 m   **6** 61.5 cm^2

7 28.7 km   **8** 21.0°   **9** 19.1 cm^2

10 69.3°   **11** 29.8 cm   **12** 25.3 cm

Chapter 23

1 a 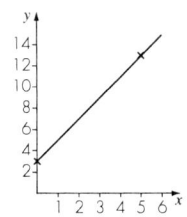   **b** 1.75   **2 a** 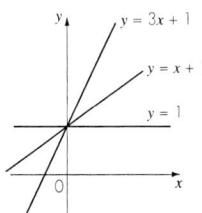   **3 a** $y = -3x + 9$ **b** −9   **c** $\frac{1}{3}$

4 a i and iv   **b** i and vi, ii and v, iv and vi   **c** iv and vi

5 $y = -2x + 5$ **6** $3y = 9 - 5x$ **7** $2y = 10 - x$

8 $y = 3x + 3$ **9** $y = 16 - 2x$ **10** (2, 1)

Chapter 24

1 a B, D, A   **b** 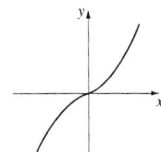   **2 a**

x	0	1	2	3	4
y	1	0.8	0.64	0.512	0.41

b   **c** 1.23

Chapter 25

1 a i 9, 5   **ii** subtract 4   **b i** 122   **ii** 20   **iii** −7

2 a Always even   **b** Could be either odd or even

3 a 2, 5   **b** 11th term   **c** (85 + 1) ÷ 3 does not give a whole number

4 a i 3, 7, 11   **ii** No, because 133 is not divisible by 4.   **b** $4n + 1$

5 a i $s^3 + 6s$ **ii** $7x - 2$ **iii** $n^2 + 6n + 9$ **b i** $(a + 0)(2a + 1)$ **ii** $4xy^2 (2x^2 - y)$

6 a

Pattern	1	2	3	4	5
No. of dots	5	8	11	14	17

b 23   **c** $3n + 2$ **d** 20th pattern

7 $x = \pm\sqrt{w - y}$

8 $x = 3$

9 b $r = \sqrt{\dfrac{A}{4 - \pi}}$

10 $x = \dfrac{Z(p - 1)}{2p}$

11 $x = \dfrac{3 + 5y}{y - 2}$

12 $x = \frac{1}{2}$, $y = 2$

13 $x = 0$ or $x = \frac{1}{6}$

14 $\dfrac{3x + 1}{x + 3}$

15 $x = 1$ or $-\frac{2}{3}$

16 $\dfrac{x - 3}{x}$

17 $x = -1\pi$

Chapter 26

1 a $x \geqslant -1$ **b** $x < 2$ **c** $-1, 0, 1$

2 a $x < 2.5$ **b** 2

3 a $y < 1.5$ **b** $r = \dfrac{p - 3}{2}$ **c** $x = 32$

4 a $x \geqslant 3$ **b i** $x > 3.6$ **ii** 4

5 $-2, -1, 0, 1, 2, 3$

6 a $x < 4$ **b** $-5 < x < 5$

7

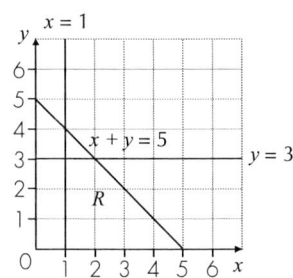

8

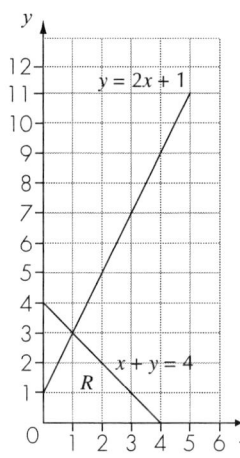

9 $x \geqslant -3$, $y \leqslant 2$, $y \geqslant x$

10

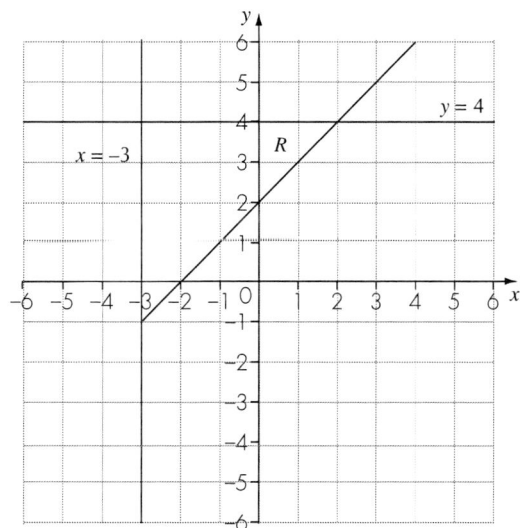

11 $-3, -2, -1, 1, 2, 3$

Chapter 27

1 Graph A: $y = (x - 3)^2$
 Graph B: $y = (x + 3)^2$
 Graph C: $y = -x^2$
 Graph D: $y = 3 - x^2$

2 **a** 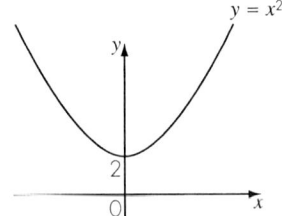 **b** 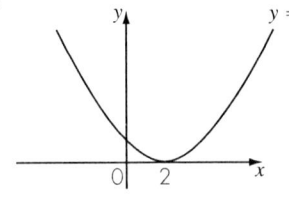 **c**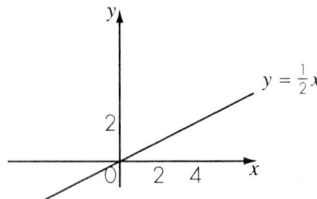

3 **a** $y = \cos x$ **b i** $y = \cos x + 1$ **ii** $y = 2 \cos x$ **iii** $y = \cos 2x$

4 **a** 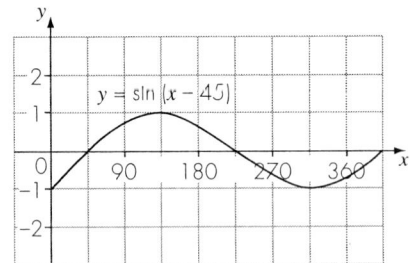 **b**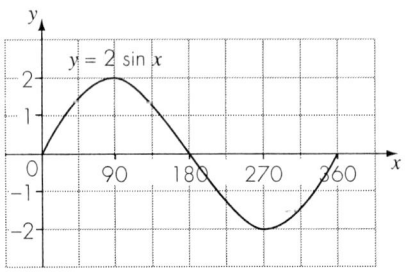

Chapter 28

1 $n + n + 1 + n + 2$
 $= 3n + 3$
 $= 3(n + 1)$

2 $6^3 - 5^3 = 91 = 13 \times 7$ which is not prime

3 Angle A = Angle C (alternate angles)
 Angle B = Angle D (alternate angles)
 AB = CD
 So, triangles are congruent (ASA)

4 **a** $n \times (n + 1) + (n + 2)^2 + (n + 1) \times (n + 4)$

 b By multiplying out brackets and simplifying, both expressions are equal to: $3n^2 + 10n + 8$